BIOCHEMISTRY RESEARCH TRENDS

TERPENOIDS AND SQUALENE

BIOSYNTHESIS, FUNCTIONS AND HEALTH IMPLICATIONS

BIOCHEMISTRY RESEARCH TRENDS

Additional books in this series can be found on Nova's website under the Series tab.

Additional e-books in this series can be found on Nova's website under the e-book tab.

BIOCHEMISTRY RESEARCH TRENDS

TERPENOIDS AND SQUALENE

BIOSYNTHESIS, FUNCTIONS AND HEALTH IMPLICATIONS

ALANNA R. BATES
EDITOR

New York

Additional color graphics may be available in the e-book version of this book.

Library of Congress Cataloging-in-Publication Data

ISBN: 978-1-63463-656-8

Library of Congress Control Number: 2014956985

Published by Nova Science Publishers, Inc. † New York

CONTENTS

PREFACE

Terpenoids represent the largest and most widely distributed class of natural, high-value chemicals, comprising more than 55,000 distinct molecules. Besides the wide range of biological properties, these natural compounds have found significant commercial value as pharmaceuticals, flavours and fragrances, commodity chemicals, and, more recently, as potential biofuels. Squalene is a natural lipid belonging to the terpenoid family and a precursor of cholesterol biosynthesis. Squalene has attracted increasing attention during the past few years due to their biological activities and natural abundance and is potential targets for the food and pharmaceutical industries. This book discusses the molecular properties, applications, functions and health implications of terpenoids and squalene.

Chapter 1 – The significant progress achieved in recent years in the emerging field of synthetic biology, combined with the use of metabolic engineering methods, has allowed access to a wide variety of natural terpenoids produced by genetically modified microbial strains. Nonetheless, improvement of microbial processes remains necessary in order to meet the volumetric productivities that are required to establish an industrially relevant production system. The physicochemical nature (e.g. low solubility or high volatility) of terpenoids and the frequent toxic effects on microbes often represent the main drawbacks encountered in microbial production. The challenge of reaching high volumetric and total yields of desired terpenoids consists in improving the overall bioprocess, using sophisticated methods to overcome limitations and to render it economically viable. The use of integrated bioprocesses by the application of in situ product recovery (ISPR) technology can provide a technical solution in the microbial manufacturing of terpenoids, with a view to optimizing yields and cost-effectiveness. This chapter aims to highlight, with selected examples, the specific challenges in the microbial production of high-value terpenoid products. The critical aspects and limitations frequently associated with terpene molecules during a microbial process are discussed. Also, technical solutions for significantly improving the overall performance of bioprocesses for high level production of valuable terpenoids by applying suitable ISPR technology are suggested.

Chapter 2 – Triterpenoids are a large group of natural compounds, many of which have interesting biological activities. Despite this, only a few derivatives are currently used as therapeutics. The main reason is that the active triterpenes usually have unfavorable pharmacological and physicochemical properties such as solubility and bioavailability. Therefore, many compounds with excellent activity in *in vitro* tests usually fail during clinical trial. Many research groups have tried to synthesize a variety of prodrugs in order to improve pharmacological properties of the active compound while retaining the desired biological activity. The most commonly chosen functional groups in triterpenes to be modified as prodrugs are hydroxyl and carboxyl. In this chapter, the authors are going to summarize all of the research that has been done in the area of triterpenoid prodrugs with the main focus on the modifications of hydroxyls and carboxyls. Commonly used types of prodrug groups (such as simple alkyl esters, other esters, ethers, amides etc.) will be shown as well as unusual prodrugs with more interesting biological and pharmacological properties (glycosides, hemi esters etc.). Their preparation will be shown as well as their chemical, biological, and pharmacological properties. In addition to this, the authors will compare each type of prodrug with the other prodrug types and will discuss their limits of use. Failures will be shown as well as successful compounds. In concluding, the authors will show prodrugs with the best properties as well as the current trends in the development of the new triterpenoid prodrugs including the most recent results from their research group.

Chapter 3 – *Launaea* Cass. is a small genus of the family Asteraceae (tribe Lactuceae, subtribe Sonchinae), comprises 54 species which 09 are presented in the flora of Algeria. Plants in the genus *Launaea* have been used ethnobotanically as bitter stomachic, for treating diarrhea, gastrointestinal tracts, as anti-inflammatory, for skin diseases, treatment of infected wounds, hepatic pains, children fever, as soporific, lactagogue, diuretic and used as insecticidal. From a chemical point of view, only ten species of the genus *Launaea* Cass. have been subjected to previous phytochemical investigation, namely, *Launaea acanthoclada, L. arborescens, L. asplenifolia,L. capitata, L.*

cassiniana, L. mucronata, L. nudicaulis, L. pinnatifida, L. resedifolia and *L. tenuiloba*. Different secondary metabolites including terpenoids, triterpenoid saponin, sesquiterpene lactones, steroids, polyphenolic compounds have been reported. On the other hand, terpenoids are chemio-characteristic of Asteraceae family, including the *Launaea* genus, have been reported to have anti-inflammatory activities, anti-hyperlipidemia, hepatoprotection, antioxidant, cytoprotective, giving protection against cardiovascular disease, and certain forms of cancer.

The authors cover in this chapter the nature of terpenoids together with their biological activities to provide a comprehensive compilation of these compounds from the genus *Launaea Cass.*, especially those growing in Algerian Sahara and used as medicinal plants, namely: *Launaea arborescens, L. nudicaulis and L. residifolia.*

Chapter 4 – Squalene is a natural lipid belonging to the terpenoid family and a precursor of cholesterol biosynthesis. Squalene has attracted increasing attention during the past few years due to their biological activities and natural abundance and is potential targets for the food and pharmaceutical industries. Here the authors summarize the recent knowledge of squalene biochemistry, its molecular properties, and its physiological effects. The analytical techniques used to identify and quantify squalene are also reviewed.

In: Terpenoids and Squalene
Editor: Alanna R. Bates
ISBN: 978-1-63463-656-8

Chapter 1

INTEGRATED BIOPROCESS APPROACHES FOR THE MICROBIAL PRODUCTION OF TERPENOIDS

Marco Antonio Mirata[1*], Hendrik Schewe[2] and Jens Schrader[3]

[1]Speciality Ingredients Research & Technology (RSI); Lonza Ltd; Switzerland

[2]DECHEMA Research Institute Biochemical Engineering; Frankfurt/Main, Germany

[3]DECHEMA Research Institute Biochemical Engineering; Frankfurt/Main, Germany

ABSTRACT

The significant progress achieved in recent years in the emerging field of synthetic biology, combined with the use of metabolic engineering methods, has allowed access to a wide variety of natural terpenoids produced by genetically modified microbial strains. Nonetheless, improvement of microbial processes remains necessary in order to meet the volumetric productivities that are required to establish an industrially relevant production system. The physicochemical nature (e.g. low solubility or high volatility) of terpenoids and the frequent toxic

[*] Corresponding author's e-mail: marco.mirata@lonza.com.

effects on microbes often represent the main drawbacks encountered in microbial production. The challenge of reaching high volumetric and total yields of desired terpenoids consists in improving the overall bioprocess, using sophisticated methods to overcome limitations and to render it economically viable. The use of integrated bioprocesses by the application of in situ product recovery (ISPR) technology can provide a technical solution in the microbial manufacturing of terpenoids, with a view to optimizing yields and cost-effectiveness. This chapter aims to highlight, with selected examples, the specific challenges in the microbial production of high-value terpenoid products. The critical aspects and limitations frequently associated with terpene molecules during a microbial process are discussed. Also, technical solutions for significantly improving the overall performance of bioprocesses for high level production of valuable terpenoids by applying suitable ISPR technology are suggested.

INTRODUCTION

Terpenoids represent the largest and most widely distributed class of natural, high-value chemicals, comprising more than 55,000 distinct molecules [1]. Besides the wide range of biological properties, these natural compounds have found significant commercial value as pharmaceuticals, flavours and fragrances, commodity chemicals, and, more recently, as potential biofuels. Over the last decades, the production of terpenoid compounds has been mainly achieved through chemical synthesis or plant extraction. For instance, to form terpene molecules by conventional synthesis, a series of isoprene condensations, cyclization and oxidation reaction are commonly used, although most of these compounds are extremely difficult to produce by chemical means [2]. Consequently, a large number of the valuable terpenoid products found on the market have traditionally been extracted from their natural sources through a laborious and expensive procedure. The blockbuster drugs artemisinin and taxol, as well as the fragrance compounds valencene and sclareol, are some examples of well-established, high-value terpenoids which over the years have been extracted from their natural sources by the pharmaceutical and the flavour and fragrance industries [3]. As conventional chemical synthesis and extraction are often highly expensive, low-yielding processes, the use of the metabolic potential of microorganisms for a sustainable biotechnological production of desirable terpene molecules has been extensively investigated over the last half-century [1]. Biotechnological processes hold major advantages for the transformation and/or production of terpenoids, as microbes offer an environmentally friendly means of efficiently converting cheap raw materials, like glucose or sucrose, into a wide variety of enzymes, providing a high regio- and stereo-specificity and a high reaction velocity even at low ambient temperature [4]. Until a few years ago, the microbial generation of valuable terpenoids was mainly bound to whole-cell or enzymatic biotransformation of cheap and abundant terpene hydrocarbons into high-value oxyfuctionalized terpenoids. *R*-(+)-limonene, *α,β*-pinene and (+)-valencene, which accumulate as waste streams of the food and wood processing industries, are examples of cheap starting materials that have been used for some time in terpene biotransformations [4]. Recently, comprehensive reviews dealing exclusively with the microbial transformation of terpenes available from renewable resources have been published by several authors [1, 4, 5]. Today, significant advances in genetic engineering over recent years have converted microbes to "microbial cell factories" for the production of a wide range of important chemical compounds. In this connection, extraordinary progress has been achieved in the engineering of microbial cells, capable at present of producing a vast variety of terpenoids that could not have been imagined a few years ago. The use of -omics technologies and bioinformatics tools for the discovery of biosynthetic enzymes, combined with the application of synthetic biology and metabolic engineering methods, has facilitated engineering of industrially suited hosts, such as *Escherichia coli* and *Saccharomyces cerevisiae*, for the production of desirable terpenoids [3, 6]. Nonetheless, the step towards a commercially exploitable manufacturing bioprocess often implies some challenges due to the low solubility, high volatility, and frequent toxic effects on microbes exhibited by this compound class. This is particularly the case for the production of terpenoids belonging to the family of the mono-, sesqui- and diterpenoids. In this compound class, a more complex structural and physicochemical variability and frequent cytotoxicity are observed in contrast to triterpenoids and carotenoids [7]. As a result, the biotechnological generation of desirable terpene compounds often results in limited productivity and diluted product streams, leading to high processing costs. In this last case, the use of sophisticated methods, such as ISPR techniques, can provide multiple benefits: to overcome product toxicity thus increasing productivity, to stabilize the product, and to facilitate further product recovery [8, 9]. The implementation of ISPR methods for the production of microbial terpenoids has in the past provided technical solutions for improving the performance and cost-effectiveness of a process. The recent progress made in engineering of microbial cells now enables the production of a huge diversity of terpene molecules. This new wave will surely lead to a growing market demand for this compound class in the near future, thus exerting increasing pressure on scientists and engineers to develop cheap and highly efficient manufacturing processes. This chapter aims to discuss the critical aspects often encountered in the microbial production of terpene products. Here, particular focus is given to bioprocesses, the limitations of which are mostly associated to the low solubility, high volatility, and frequent toxic effects on microbes of the terpenoid compounds. Also, suggestions and guidelines are highlighted to improve the overall performance of microbial processes used to produce high-value terpenoids by the application of ISPR technology.

TERPENOIDS: OCCURRENCE, PHYSICOCHEMICAL PROPERTIES AND CYTOTOXICITY

Terpenoids, also called isoprenoids, are natural compounds that are ubiquitous in all classes of living organisms, forming the largest and most widespread family of metabolites, the majority of which occur in plants. All terpenoids originate from the universal C_5 precursors isopentenyl pyrophosphate (IPP) and its isomer dimethylallyl di-phosphate (DMAPP). One of two possible natural routes, the mevalonate pathway (MVA) or the methylerythritol pathway (MEP), contributes to the formation of the C_5 ("isoprene") units, which are in turn the precursors of numerous isoprenoid molecules made by the terpene synthase enzymes, which are further modified by the actions of various enzyme classes. Isoprenoids are classified based on the number of isoprene units linked, such as hemiterpenes (C_5), monoterpenes (C_{10}), sesquiterpenes (C_{15}), diterpenes (C_{20}), triterpenes (C_{30}) or carotenoids (C_{40}). Most of these compounds, which accumulate at very low concentration, are of key importance in nature because of their varied biological functions [10]. Terpenoids encompass a huge diversity of molecular structures that can be more or less complex, resulting in a broad variability of physicochemical properties (i.e.: lipophilicity, size and chirality, boiling point, degree of volatilization, etc.). For example, the mono- and sesquiterpenes that count for most abundant terpenes, mostly found in plant essential oils, readily volatilize from an aqueous medium. Despite its high boiling point of 176°C, the volatilization of the monoterpene hydrocarbon limonene from a water surface is expected to occur rapidly based upon the estimated low Henry's Law constant $\kappa_{H,cp}$((0.048 mol L^{-1} atm^{-1} at 25°C). Henry's Law, describing the solubility behaviour of volatile compounds in a liquid, says that at low concentrations the partial pressure of the volatile substance in the gas phase is directly proportional to its solubility in the liquid phase. Given that the vapour pressure of limonene (0.0019 atm at 25°C) is high and its solubility in water is low (13.8 mg L^{-1} at 25°C), a high rate of vaporization is therefore predicted [11, 12]. Back in 2001, limonene bioconversion to the antimicrobial compound perillic acid was carried out in a fed-batch aerated bioreactor using a growing *Pseudomonas putida* DSM 12264 culture. To prevent loss of the hydrophobic and volatile precursor via stripping, the airflow was saturated with pure limonene prior to addition to the culture, whereas glycerol was fed as a C-source [13]. In another example, Newman et al. (2006) confirmed experimentally that volatilization of amorpha-4,11-diene, a key intermediate in the semi-synthesis of the anti-malarial drug artemisinin, was the primary mechanism for the loss of this sesquiterpene during fermentation using an engineered *E. coli* strain, rather than microbial degradation. The monitoring of amorpha-4,11-diene volatilization from the bioreactor under standard conditions without cells showed first order decay with a less than one hour half-life, thus confirming that amorpha-4,11-diene loss occurs rapidly from a clarified medium. This high volatilization rate was expected, due to the fact that amorpha-4,11-diene is sparingly soluble in water and in an immiscible liquid mixture, the vapour phase is enriched for the diluted component of the system [14]. Oxygenated terpenes are usually less volatile than their corresponding hydrocarbons as they exhibit higher water solubility and lower vapour pressure [11]. For instance, linalool, a naturally occurring monoterpene alcohol found in many flowers and spicy plants, is less readily to volatilize than limonene from an aqueous system based upon a higher Henry's Law constant $\kappa_{H,cp}$ (53 mol L^{-1} atm^{-1} at 25°C) calculated from its water solubility (1.58 g L^{-1} at 25°C) and vapour pressure (0.0002 atm, at 25°C) values. In a study carried out by Wriessnegger et al. (2014), the evaporation of the sesquiterpene hydrocarbon (+)-valencene, used as a precursor for the production of the grapefruit aroma (+)-nootkatone, was presumed one of the major limitations during a biooxidation process using an engineered *Pichia pastoris* yeast as whole-cell biocatalyst. Due to the lower solubility of (+)-valencene as compared with the oxygenated intermediate and target products *trans*-nootkatol and (+)-nootkatone in an aqueous environment, a moderate conversion yield was observed due to the high evaporation of this sesquiterpene hydrocarbon substrate during the biotransformation. In contrast, (+)-nootkatone formation was improved considerably when the intermediate *trans*-nootkatol was added as sole precursor molecule, whereas losses of the substrate and of the product were minimal after 24 hours' biotransformation in this last experiment [15]. These few examples demonstrate that basic data on the physical properties of terpenoids are useful for developing a microbial process, given that in pure aqueous systems, yield as well as loss of substrate and product will be directly affected by the low solubility and the high volatility of terpene molecules. Hereby, the octanol-water partition coefficient log $P_{octanol/water}$ can be used to easily estimate the water solubility of chemicals with lipophilic and hydrophobic characteristics such as terpenoids. Log *P* is the ratio of a chemical's concentration in the octanol phase to its concentration in the aqueous phase of a two-liquid-phase system at the equilibrium. For example, in the study carried out by Cal in 2006, the linear correlation between log *P* and the water solubility of seven hydrocarbon-, ester- and alcohol-type terpenes could be experimentally confirmed by investigating the relationship between the experimental aqueous solubility data of the tested compounds and their log *P* values calculated by ADC/Log *P* software [16]. Many other studies have shown that physical properties such as water solubility or log *P* are also valuable parameters for correlating the tendency of terpene compounds to interact with biological membranes. Besides the problems arising from low aqueous solubility and volatility, the frequent toxic effects of terpene molecules on microorganisms can be another significant hurdle during bioprocess development [17]. Owing to their hydrophobicity, terpene compounds preferentially partition to lipid structures, which makes the cellular membrane the main target for product accumulation during microbial processes. Mechanisms of action of cytotoxic terpene molecules in microbes are not well understood. For hydrophobic compounds such as terpenes that are often slightly soluble in the aqueous phase, their main action of toxicity to microbes can be expected to be on the molecular level. Due to their lipophilicity, these compounds intercalate within the lipid bilayer and deteriorate membrane function. Increase in membrane fluidity, membrane damage, denaturation of membrane-bound proteins, and loss of energy transduction are some of the consequences of a solvent's impact at the molecular level. When the amount of an organic solvent such as a terpene exceeds its maximum solubility in the culture medium and a distinct second phase exists, additional phase toxicity may occur. The presence of a second phase means that thermodynamically, the aqueous phase is fully saturated, as well as the cells and their different compartments (e.g. cytoplasmic membrane). Phase toxicity is probably caused by extraction of outer-cellular components during cell-solvent contact, extraction of nutrients from the media, or cell coating [18]. Several parameters have been studied to describe the harmful effect of solvents such as terpenes on microbes. Hereby, the log *P* is also an ideal parameter to measure the toxicity of a terpene. The greater the polarity of the solvent, the lower its log *P* value and the greater its toxicity. In general, solvents with log *P* values between 1 and 4.5 are considered extremely toxic to microorganisms, whereas strong hydrophobic solvents with log *P* greater than 5 are commonly used as a non-toxic second phase that functions as a reservoir for toxic compounds in two-liquid-phase partitioning bioprocesses [9, 18]. An empirical correlation was found between the log $P_{octanol/water}$ value of an organic solvent and its partitioning in a cell membrane-water system (log $P_{membrane/water}$) [19].

$$\text{Log } P_{membrane/water} = 0.97 \log P_{octanol/water} - 0.64$$

With this equation, the actual membrane concentration of a hydrophobic compound can be estimated if its concentration in the water phase is known. For instance, linalool (log $P_{octanol/water}$ value of 2.97) will accumulate within the membranes in concentrations of up to 1.7 M if it is present in the water phase at a saturation concentration of only 10 mM. This concentration would clearly not allow conventional microorganisms to survive. In a study carried out in 2008, linalool toxicity was determined experimentally for seven different fungal strains to ensure a reasonable precursor supply during fed-batch cultivation experiments while avoiding any toxic effect. Data obtained from this toxicity study were used to grow the selected microorganisms with strain-specific linalool feedings at the threshold concentration (1mM) where a sharp decrease in the cell viability would occur. By these means, extended biomass growth and higher linalool biotransformation product concentrations were targeted. This substrate feeding approach enabled the identification of the novel linalool bioconversion products lilac aldehyde and lilac alcohol, characteristic fragrance compounds found in the lilac flower [20]. In another example, a detailed product inhibition study of the biotransformation of *R*-(+)-limonene to *R*-(+)-perillic acid by *P. putida* DSM 12264 revealed that at elevated concentration (164 mM), the monoterpenoic acid inhibits growth as well as the biotransformation activity of the biocatalyst, whereas this strain is capable of growing with a neat phase of the hydrophobic limonene. Despite the extraordinarily high tolerance of *P. putida* towards toxic compounds, these inhibition effects can be attributed to the good aqueous solubility of perillic acid at a neutral pH and to its relatively low log *P* value (2.41). The data obtained from the product inhibition studies clearly illustrated that keeping aqueous perillic acid concentration at a low level was an essential prerequisite for increasing productivity and maximum perillic acid concentration. These findings were the foundation for the development of an anion exchange-based in situ product recovery system coupled to a fed-batch bioreactor that allowed a significant improvement of the bioproduction of perillic acid from limonene with *P. putida* DSM 12264 [21]. More recently, concerns about limited fuel resources and rising petroleum costs have stimulated a growing interest in the rapid engineering of novel terpene-based microbial platforms for the production of advanced biofuels. Nonetheless, the feasibility of fermentative production of isoprenoids using engineered microorganisms as a potential source of biofuels is primarily dependent on the toxicity it imparts to the overproducing organism, in addition to a high microbial titer. Such a consideration has become essential in the work of Peralta-Yaha et al., who in 2011 identified a novel biosynthetic alternative to D2 diesel fuel, bisabolane, and engineered microbial platforms for the production of its intermediate precursor, the sesquiterpene bisabolene. Investigations on the toxic effects of this sesquiterpene on the overproducing *E. coli* and *S. cerevisiae* strains were necessary to evaluate the feasibility of bisabolene as a biofuel precursor. Growth experiments with above 20% (v/v) bisabolene added exogenously evidenced a lack of toxicity in this sesquiterpene to the tested microbial platforms. Thanks to these investigations, it could be proved that production of bisabolene at very high titers is in principle possible, which is one of the major conditions to consider in the development of a manufacturing process [22]. In this field, farnesene is probably one of the most famous examples of an advanced sesquiterpene-based biofuel introduced to the market as an alternative biodiesel and jet fuel, recently produced at very high titers by yeast fermentation for large-scale industrial production [18, 23].

In Situ Product Removal: Principles and Application in Microbial Production of Terpenoids

Experience shows that a basic knowledge of the physicochemical properties and the cytotoxicity of terpene compounds involved in a microbial production process become essential at an early development stage. Problems resulting from cytotoxicity, as well as from undesired loss of terpene molecules occurring during cultivation, can thus be reduced or even avoided using an integrated bioprocess. A variety of process engineering solutions exist for this purpose, such as fed-batch cultivation by applying precursor feeding or by designing effective in situ product removal (ISPR) technology. ISPR represents a tool of choice especially for those processes which are limited by unstable and/or toxic compounds. In particular, it can also facilitate further downstream processing with direct and selective removal of the product during the growth and production phases [4]. Bearing in mind the fact that in general, high productivities are expected to make a bioprocess cost competitive, the application of an ISPR approach in the production of microbial terpenoids will contribute to reducing the overall processing costs by increasing volumetric productivity and total yields and by improving the final product quality. Over the past few decades, numerous examples of successful applications of ISPR technologies have been reported for the recovery of different types of molecules such as flavours and fragrances, alcohols, organic acids, organic solvents, secondary metabolites, and proteins, and extensive reviews of ISPR projects and process designs have been done [9, 24, 25]. The implementation of a suitable separation technique in a bioprocess implies the selection of a proper ISPR strategy (recovery method), of the most appropriate process configuration (e.g. internal or external ISPR), and of the best mode of operation (e.g. fed-batch or continuous process). Here, the development of a structured approach, where the characteristics and constraints of a microbial process are rigorously evaluated, will assist key decisions on the design and selection of an effective and reliable integrated bioprocess. The separation principles used for the ISPR are based on the exploitation of the existing difference with respect to a certain physical or chemical property between the target substance and the background of the medium. In this regard, discrepancies in the relative volatility, hydrophobicity, size, charge, solubility in a second phase, and specific binding are the key features that are considered when assigning an appropriate ISPR methodology for a product to be separated from a biological system [26, 27].

Over recent decades, ISPR techniques have advanced significantly and their potential for improving yield and productivity has been demonstrated for several processes. ISPR comprises the following techniques: evaporation via vacuum, gas stripping, adsorption on macroporous resins, pervaporation, liquid-liquid extraction, ion exchange, crystallisation, and electrodialysis [9]. Heuristics for the selection of proper ISPR technology are shown in Figure 1.

The success of an integrated bioprocess depends not only on the selected separation technology, but also on the configuration of the bioreactor/ separation units and the mode of operation.

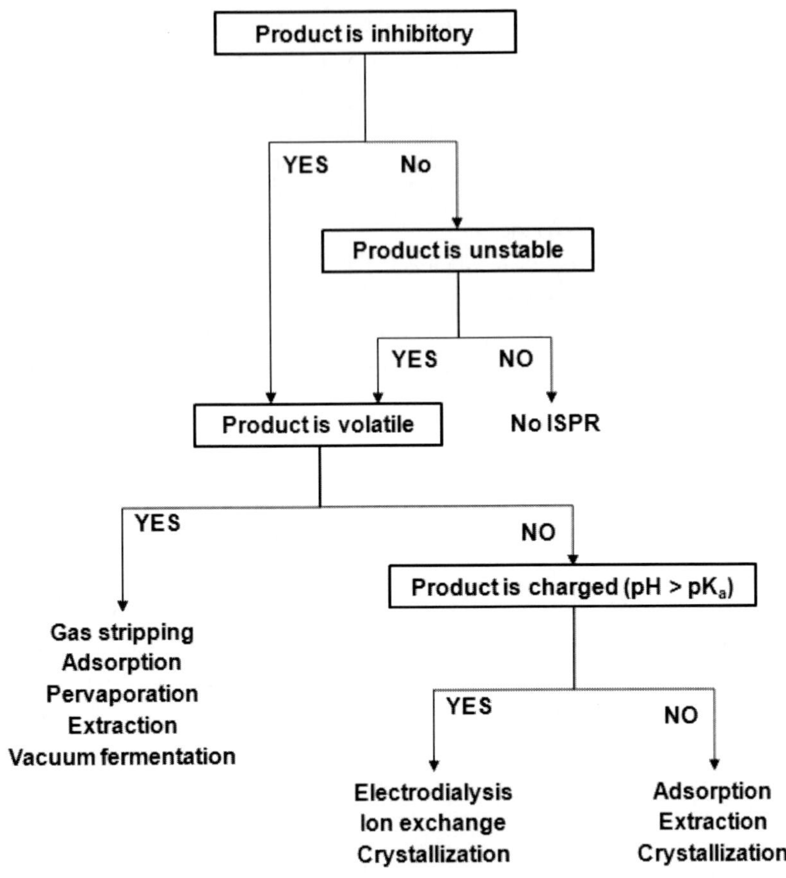

Figure 1. Heuristics for the selection of the appropriate ISPR technique [25].

Several possible ISPR process configurations emerge, whether the separation of cells and product can be achieved by employing an *internal configuration*, where the fermentation liquid is in contact with a product-removing phase that is present in the fermenter; or an *external configuration*, where the fermenter liquid is in contact with a product-removing phase in an external unit via a loop. In either the internal or external configuration, the cells can be in *direct* or *indirect* contact with the product-removing phase [24, 26, 28]. The choice of configuration will ultimately depend on the properties of the target product and of the biological system, the separation method, and the mode of operation. The basic configurations for ISPR as applied to the established techniques are shown in Figure 2.

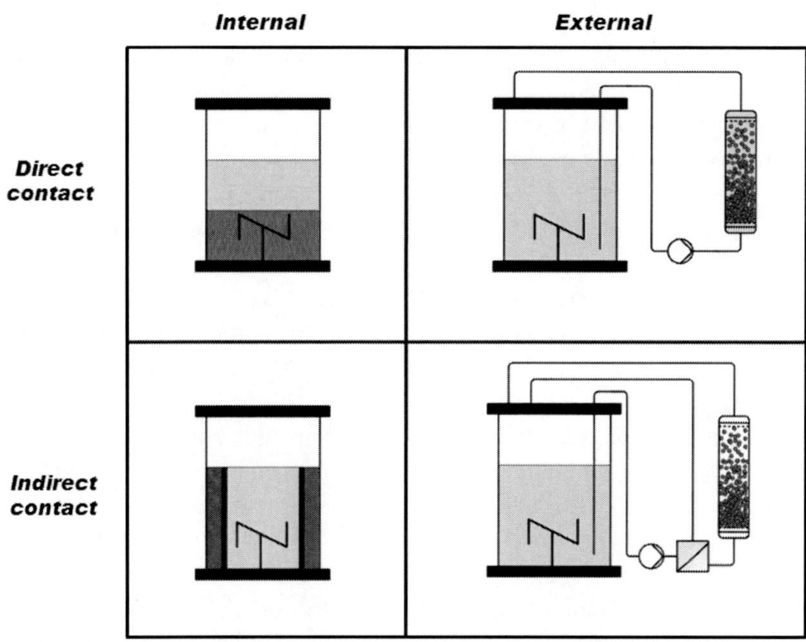

Figure 2. Process configuration for in situ product recovery. Separation may take place inside the reactor (internal) or in an external unit. In some cases direct contact between microorganisms and extracting phase may take place. In other cases contact is avoided by means of some cell separation system (indirect contact) [28].

Most investigated ISPR techniques used to alleviate problems arising in microbial processes from low water solubility, high volatility and cytotoxicity of terpenoids are liquid-liquid extraction, adsorption, and gas stripping. Hitherto, ISPR has mainly been applied in the microbial production of mono-, sesqui- and diterpenoids because a more complex structural and physicochemical variability and frequent cytotoxicity are observed in this class of compounds in comparison with triterpenoids and carotenoids [7]. For this reason, this chapter essentially focuses on examples associated with the biotechnological production of mono-, sesqui- and diterpenoids by using either whole-cell biotransformation or fermentation with engineered microbial hosts (see Table 1). So far, liquid-liquid extraction has shown the greater potential because of its simplicity of application. Several examples of biotransformation involving terpenes have been reported as having enhanced the production of their oxyfunctionalized derived products in two-liquid-phase partitioning bioreactors, where a water-immiscible organic solvent is used as an extracting phase to avoid toxic or inhibitory effects on the biocatalysts (see Figure 2, internal configuration with direct contact). Both precursor and product are separated and stored in a non-toxic hydrophobic phase from the aqueous phase, where biocatalysis takes place [4]. In an effort to improve the microbial transformation of (−)-*trans*-carveol to the flavour and fragrance compound (*R*)-(−)-carvone by *Rhodococcus erythropolis* DCL14, Morrish et al. (2008) followed a rational solvent regime selection to develop a two-liquid-phase partitioning. In this work, 1-dodecene was selected as an appropriate solvent based on several factors including biocompatibility, biodegradability and partitioning capacity. The use of this solvent for an in situ extraction of the toxic substrate (*cis*-carveol) and product (carvone) during the biotransformation thus allowed to minimize inhibition effects induced by *cis*-carveol and carvone [29]. In general, a biocompatible solvent ensures that its presence will not be inhibitory to the organism or otherwise affect its metabolic activity. Depending on the microorganism and the molecular structure of the organic solvent used, organic solvents with log *P* values higher than 5 are generally recognized as biocompatible [30]. Biodegradability is another very important factor in solvent selection. It is important that the organism does not metabolize the solvent, as this can cause loss of solvent and production of undesired by-products and can also lead to substrate competition, where the solvent is preferentially metabolized relative to the desired substrate. A high partition coefficient, calculated as the ratio of the product concentration in the organic phase to its concentration in aqueous, is another important requirement for the selection of a good extraction solvent. Safety and cost considerations are other key criteria that will leverage the choice of a suitable solvent [29].

To address low substrate solubility and toxicity issues, a rational solvent selection approach was also followed by Schewe et al. (2008) in the development of an aqueous two-liquid-phase bioprocess for the oxidation of the bicyclic monoterpene α-pinene to α-pinene oxide, verbenol, and myrtenol using an engineered *E. coli* overexpressing a P450BM-3 QM. In this study, the selection of a biocompatible organic carrier solvent with favourable partition coefficients was crucial for the implementation of an effective bioconversion in an aqueous–organic biphasic system. Out of seven organic solvents tested, diisononyl phthalate (log *P* value of 8.8) was selected as the most biocompatible organic carrier solvent, capable of masking the toxic effects mediated by α-pinene and of efficiently extracting the products. Applying this engineered *E. coli* strain in a two-liquid-phase partitioning bioprocess (phase ratio 2:3 (v_{org}:v_{tot}); organic phase consisted of 70% (v/v) DINP and 30% (v/v) α-pinene) a total product concentration of over 1 g L^{-1} of oxidized products could be achieved after only four hours, which are the highest data reported so far for a microbial α-pinene oxyfunctionalization [30]. More recently, Cornelissen et al. (2013) established a two-liquid-phase biotransformation for the hydroxylation of (*S*)-limonene to the anticancerogenic compound (*S*)-perillyl alcohol using an engineered *E. coli* strain harbouring a CYP153A6 monooxygenase and the outer membrane protein AlkL. Here, an initial twofold increase of limonene hydroxylation activity was achieved in a single-phase system by co-expressing CYP153A6 and AlkL genes in *E. coli*, to enable an improved substrate uptake via the outer membrane protein AlkL. The application of the same biotransformation in a two-liquid-phase system, with bis(2-ethyl-hexyl) phtalate (log *P* value of 7.5; phase ratio 1:3 (v_{org}:v_{tot})) as an organic carrier solvent to prevent toxic effects on substrate and product and to enable in situ product extraction, enhanced hydroxylation activity fivefold. In the latter system, the *E. coli* strain harbouring CYP156A6 and AlkL produced almost 6 g L^{-1} perillyl alcohol within 26 hours [31]. Besides their use in whole-cell biotransformation, two-liquid-phase partitioning bioreactors have been successfully applied during the last few years in many processes using engineered microbial strains for the production of value-added terpenoids that may be lost from fermentations due to air-stripping or may cause cytotoxic effects to the cells. The microbial production of the anti-malarial drug precursor amorpha-4,11-diene is a pioneering example of terpenoid biosynthesis using engineered microorganisms. In 2006, Newman et al. demonstrated that volatilization is the primary mechanism for loss of amorpha-4,11-diene from the clarified medium during the cultivation of an engineered *E. coli* strain. To circumvent this problem, an in situ extraction of the released sesquiterpene was performed in a two-liquid-phase partitioning bioreactor using an organic layer of dodecane (log *P* of value 6.6; 10% (v/v)) as a biocompatible secondary phase. As a result, the amorpha-4,11-diene produced from the cell quickly partitioned to the dodecane phase. This caused amorpha-4,11-diene to become trapped in the hydrophobic layer and to improve titer twentyfold, generating approximately 0.5 g L^{-1}[14]. By combining strain and process improvements, the same research group later achieved commercially relevant titers with the production of 27.4 g L^{-1} amorpha-4,11-diene with a superior two-liquid-phase fermentation process using a further optimized *E. coli* strain [32].

An impressive effort to develop an efficient route to artemisinic acid in an engineered *S. cerevisiae* strain, able to overproduce and oxidize amorpha-4,11-diene by extractive fermentation using a feedback controlled pulse process, was recently demonstrated by Paddon et al (2013). In this process, 10% (v/v) isopropyl myristate oil (IPM) was added to the cultivation medium, which allowed effective in situ extraction of artemisinic acid. As IPM (log *P* value of 7.43) is a biocompatible solvent, the in situ extraction procedure reduced toxicity effects on the *S. cerevisiae* producing strain, thereby resulting to a titer of 25 g L^{-1} artemisinic acid. Furthemore, a practical, efficient and scalable chemical process for the conversion of artemisinic acid to artemisinin was developed. This semi-synthetic process builds on the fundaments of a viable industrial process to increase world provision of first-line antimalarial treatments [33]. In an effort to find an alternative route to conventional plant extraction of the diterpene sclareol, Schalk et al. (2012) were recently able to synthetize this key fragrance intermediate in a genetically optimized *E. coli* strain at a concentration of 1.5 g L^{-1}, using a two-liquid-phase partitioning bioprocess. Similarly to the previous example, a 10% (v/v) dodecane overlay was added to the high cell density fermentation to trap the product in the organic phase, thus avoiding losses via stripping. This is the highest titer reported so far for a diterpene production in a microbial host, showing as well an attractive route to conventional sclareol plant extraction [34]. Recently, Wriessneger et al. (2014) created a metabolically engineered *Pichia pastoris* strain that was able to produce industrially interesting quantities of (+)-nootkatone, upon cultivation in a two-liquid-phase partitioning bioreactor with a dodecane overlay (10% (v/v)) for trapping this highly valued grapefruit aroma compound. Initially, the authors developed a whole-cell biocatalyst derived from a *Pichia pastoris* yeast able to oxidize exogenously added (+)-valencene into (+)-nootkatone. Unfortunately, (+)-nootkatone formation using this whole-cell biocatalyst was seriously limited due to loss of (+)-valencene by evaporation. On the other hand, the application of a biphysic system was not a good option, as (+)-valencene was quantitatively trapped in the organic solvent and neither evaporation nor conversion by the cells resulted in appreciable product amounts. To resolve the problems caused by substrate losses and phase-transfer issues, a straightforward strategy was then established to produce intracellular (+)-valencene by co-expression of a valencene synthase in the *P. pastoris* whole-cell biocatalyst. The last improvements introduced to the microbial host allowed for significant increase of (+)-nootkatone yield to 208 mg L^{-1} cell culture in a two-liquid-phase partitioning bioreactor [15]. Currently, engineering biosynthetic pathways for monoterpene production in microbial hosts as a potential route for delivering petroleum-compatible fuels is receiving increasing attention. However, unlike farnesene (log *P* value of 6.14) and bisabolene (log *P* value of 7.80), monoterpenes are generally highly toxic to industrially suited microorganisms, including *Saccharomyces cerevisiae* and *Escherichia coli*, due to log *P* values that are typically lower than 5. Hereto, the realization of microbial monoterpene production will rely extensively on solving this core issue. To address this problem, Willrodt et al. (2014) recently defined and assessed engineering strategies for the intensification of fermentative limonene production in *E. coli*. In their study, the biosynthesis of this monoterpene was optimized by systematically addressing the limitations of the microbial system associated with genetics, physiology, and reaction engineering. Amongst the parameters investigated, reaction engineering improvements, where a two-liquid-phase system was set up to produce limonene in *E. coli*, appeared to be a mandatory requirement to prevent the product evaporating and to reduce toxic effects on the host cell. Consequently, the application of a two-liquid-phase fed-batch fermentation for the biosynthesis of limonene using diisononyl phthalate as an organic carrier solvent yielded up to 2.7 g L^{-1} product in the organic fraction [35].

In another example, a proof-of-concept of a two-liquid-phase extractive system was evaluated for monoterpene production with *S. cerevisiae* to alleviate toxicity during fermentation. First, the minimum inhibitory concentration (MIC) for *S. cerevisiae* was determined for five monoterpenes comprising β-pinene, limonene, myrcene, γ-terpinene, and terpinolene. Given the low MIC for all compounds tested, ten organic solvents were then screened for biocompatibility, monoterpene distribution, phase separation, and price to select suitable solvent candidates to be used for extractive fermentation. As a result, the authors demonstrated that, using this biphasic approach, *S. cerevisiae* was able to withstand significantly higher monoterpene concentrations compared to the solvent-free system. Although farnesene did not come out as the best option for reducing monoterpene toxicity, it was of particular interest as an extractive solvent, since its use as a co-product and extractant for microbial terpene-based biofuel production in a two-phase system offers an attractive bioprocessing option for the creation of blends exhibiting physicochemical properties similar to jet fuels [18].

In situ product recovery of lipophilic compounds from aqueous media and target molecules which carry a charge at fermentation pH using hydrophobic adsorbent and ion-exchange resins was proven to be a successful tool in the microbial production of flavours and fragrances, secondary metabolites, proteins, commodity chemicals such as lactic acid, and second-generation biofuels such as butanol [36]. The use of adsorbent and ion-exchange resins offers distinct advantages among other ISPR alternatives, including: greater phase stability (physically, chemically, and biologically), improved biocompatibility, complete immiscibility with the fermentation medium, elimination of emulsification, an increased potential for re-use, and easy separation and compartmentalization from a biological system [37]. The beneficial use of an ion-exchange-based ISPR for the production of a valuable terpenoid compound has been demonstrated by Schrader and Mirata (2009) in a study on the improvement of perillic acid bioproduction by *P. putida* DSM 12264 from the monoterpene limonene. The exploitation of the carboxy group to selectively adsorb perillic acid on anion exchangers turned out to be the most promising method of alleviating product inhibition effects during the biotransformation. From seven anion-exchange resins screened, Amberlite IRA 410 Cl emerged as the most suitable resin due to its good biocompatibility with *P. putida* and high adsorption capacity for perillate anions. However, the carbon source, glycerol, and other nutrients present in the medium were found to not adsorb on the resin, in contrast with the hydrophobic substrate limonene, which exhibited a strong affinity for the polystyrene matrix of the resin.

This last observation revealed the necessity of applying a fed-batch feeding strategy, using a limonene-saturated air stream and limonene-saturated resin to guarantee a sufficient substrate supply. Subsequently, this ISPR technique via adsorption was implemented to a fed-batch bioreactor by coupling a fluidized bed of the IRA 410 Cl resin by an external recovery loop (see Figure 2, external configuration with direct contact). The continuous removal of perillic acid by recirculation of the unfiltrated broth through the ISPR unit led to a total production concentration of 31 g L^{-1} after about seven days of cultivation. The product yield achieved in that study was 2.8 times higher compared with the conventional fed-batch bioprocess without ISPR. Furthermore, the high selectivity of the ISPR method used allowed for significantly simplified downstream processing, which resulted in a highly pure product with negligible loss [38]. Knowing that perillic acid exhibits a broad antimicrobial activity and is considered a promising natural alternative to conventional preservatives, such as formaldehyde derivatives, for applications in cosmetics and pharmaceutical industries, this anion-exchange-based ISPR bioprocess represents an interesting option for industrial production of this monoterpenoic acid from cheap and renewable raw material [5]. A few years later, the same research group developed an integrated bioprocess for the stereospecific production of the fragrance compounds linalool oxides from linalool with *Corynespora cassiicola* DSM 62475. Since even low substrate and product concentrations (200 mg L^{-1}) inhibited the growth of the biocatalyst, a substrate feeding product removal (SFPR) system based on hydrophobic adsorbers was established. Owing to the similar physicochemical properties of linalool and linalool oxide (log *P* values of 3.5 and 2.4 respectively), the SFPR approach turned out to be the most favourable method of avoiding toxic effect and supplying sufficient substrate during linalool biotransformation. Amongst five screened adsorbers, Lewatit VP OC 1163 exhibited the best adsorption characteristics either for linalool or for linalool oxide, enabling a high load of the resin with the substrate while maintaining low aqueous concentration for these two monoterpenoids. Commonly, the effect expected with such a SFPR system is that the loaded substrate desorbs into the aqueous phase until equilibrium is established, thus providing a constant substrate supply according to the metabolic demand of the biocatalyst combined with instantaneous removal of the toxic product. The application of this liquid-solid two-phase SFPR technique in a bioreactor system enabled considerable reduction of the toxic effects occurring during the production of linalool oxides from linalool. Hereafter, an overall product concentration of 4.6 g L^{-1} linalool oxide was obtained, which represents a 4.6-fold increase in productivity compared with the fed-batch process without SFPR [39]. Recently, the application of an adsorption-based ISPR method was investigated in the production of perillyl alcohol from simple sugars, using an engineered *E. coli* strain, co-expressing a limonene pathway and a cytochrome P450 that specifically hydroxylates limonene in the C7 position. The presence during the biosynthesis of the resin Amberlite IRA 410 Cl, which exhibited favourable characteristics for relieving toxic effects on the product due to its stronger affinity for perillyl alcohol over limonene, enable the production of 435 mg L^{-1} limonene and 105 mg L^{-1} perillyl alcohol, respectively. Nonetheless, this study showed that the titer of perillyl alcohol was not as high as the level obtained without ISPR, although the specific production in the presence of the Amberlite resin was 1.5-fold higher than in the absence of resin [40].

In the case of volatile products, gas stripping is the most favourable ISPR method. Due to its simple mode of operation, non-harmful impact to the culture and low capital investment for facilities, this method is one of the most investigated ISPR technologies, especially in acetone–butanol–ethanol (ABE) fermentation [41]. The increasing global need of isoprene, a key commodity chemical used to manufacture a wide range of industrial products, has recently encouraged industry to invest in industrial biotechnology for the development of a gas-phase bioprocess for isoprene-monomer production. Since the yield of isoprene from naturally occurring organisms is commercially unattractive,

Genencor (now DuPont) and Goodyear developed hereto a genetically engineered *E. coli* strain for the overproduction of isoprene through fermentation of glucose. Thanks to its low boiling point (34°C) and water solubility, the isoprene emitted during the fermentation was easily recovered from the bioreactor off-gas and collected with a high purity. This gas-phase bioprocess enabled production of isoprene at a titer exceeding 60 g L^{-1}. This joint development work resulted in the production, recovery, polymerization and manufacture of prototype rubber tyres using bio-based isoprene [42].

Table 1. Selected microbial process examples using whole cell biocatalysts or engineered cell systems for the production of terpenoids performed by applying ISPR technologies

Strain	**Substrate ; C-source**	**Product**	**ISPR method; solvent or adsorbent type**	**Product titer [g L^{-1}]**	**References**
R. erythropolis DCL14	(*E*)-Carveol; Glucose	(*R*)-carvone	AOBS; 1-dodecene	0.32	[29]
Recombinant *E. coli*	α-Pinene; Glucose	α-Pinene oxide, Verbenol, Myrtenol	AOBS; Diisononyl phtalate	1.0	[30]
Recombinant *E. coli*	(*S*)-Limonene; Glucose	(*S*)-Perillyl alcohol	AOBS; Bis(2-ethyl-hexyl) phtalate	6.0	[31]
Recombinant *E. coli*	Glycerol	Amorpha-4,11-diene	AOBS; Dodecane	0.5	[14]
Recombinant *E. coli*	Glucose	Amorpha-4,11-diene	AOBS; Dodecane	27.4	[32]
Engineered *S. cerevisiae*	Glucose, Ethanol	Artemisinic acid	AOBS; Isopropyl myristate	25	[33]
Engineered *E. coli*	Glycerol	Sclareol	AOBS; Dodecane	1.5	[34]
Engineered *P. pastoris*	Glucose, Methanol	(+)-Nootkatone	AOBS; Dodecane	0.21	[15]
Engineered *E- coli*	Glycerol	(*S*)-Limonene	AOBS; Diisononyl phtalate	2.7[a]	[35]
P. putida DSM 12264	Glycerol; (*R*)-Limonene	(*R*)-perillic acid	Adsorption; IRA 410Cl anion exchanger	31	[21]
C. cassiicola DSM 62475	Glucose; (*R*)-(−)-linalool	Linalool oxides	Adsorption (SFPR); Lewatit VP OC 1163	4.6	[39]

Strain	Substrate ; C-source	Product	ISPR method; solvent or adsorbent type	Product titer [g L-1]	References
Engineered *E. coli*	Glucose	(*S*)-Limonene, (*S*)-Perillyl alcohol	Adsorption; IRA 410Cl anion exchanger	0.44, 0.11	[40]
Engineered *E. coli*	Glucose	Isoprene	Gas stripping	60	[42]

ISPR: *In situ* product recovery; AOBS: aqueous-organic biphasic system.
[a] Value calculated from organic phase volume.

FUTURE DIRECTIONS

Recent works indicate that ISPR is a well established operation in microbial production of terpenoids to overcome constraints caused by toxic and/or unstable products that often limit industrial application of such bioprocesses. In this case, the introduction of ISPR in a bioprocess is an essential measure to achieve enhancements in terms of overall product titers, productivity and yields, and to reduce processing steps [25]. Nevertheless, the immense potential of using ISPR technologies in the biotechnological production of terpenoids hasn't been fully exploited. Due to its simplicity of application, liquid-liquid extraction using organic solvent has been so far the most applied ISPR method in research (cf. Table 1). Hereto, the overall costs to manufacture attractive terpene compounds will result from a trade-off between the degree of selectivity of the ISPR method and the complexity of the up and downstream processing steps. For this purpose, the need for more sophisticated methods to remove the target terpene molecules in a highly selective manner from the biological background of the medium is essential. In future, alternative ISPR technologies, such a membrane-based methods (e.g. pervaporation and pertraction), electrodialysis, as well as the use of novel high selective and fuctional extractants (e.g. functionalized polymeric adsorbents or ionic liquids) should receive more consideration by process engineers with a view to develop more cost-effective bioprocesses [8, 36]. Although most studies where ISPR was applied for the microbial production of terpenoids have used batch or fed-batch operation systems, the set-up of ISPR in continuous cultivation mode as alternative bio-based process operation has only been rarely investigated so far [43]. On the other hand, the benefits of applying ISPR techniques are often restricted due to the physiological limitations of the well established terpene production systems as *E. coli* and

S. cerevisiae (cf. Table 1) that remain in the laboratory standard because of the established molecular biology tools and the ease of cultivation. Hereto, the development of alternative cell systems (e.g. *P. pastoris*, *Bacillus subtilis*, *Pseupomonas* sp., *P. putida*), which exhibit improved capability to express heterologous proteins and/or feature a high tolerance towards toxic compounds should be more exploited with a view of pushing forward the progress in term of product titer and yield of high-value terpenoids [7]. Therefore, it can be assumed that in the next future new and industrially attractive integrated bioprocess systems for the production of attractive terpenoids will be developed at the crossroads between metabolic engineering and bioprocess engineering, where ISPR technology will play a key role.

Conclusion

Decades of intensive research on production of terpenoids using whole-cell biotransformation has proved that ISPR technology is the tool of choice to significantly improve product titers in bioprocesses suffering from limited productivity induced by the physicochemical nature (e.g. low solubility or high volatility) and the frequent toxicity effects of terpenoids. Also, extraordinary progress achieved recently in the production of a fruitful amount of novel terpenoids using engineered cell systems, at levels that could not even imagined a few years ago, has only been possible by combining metabolic engineering approaches with ISPR technologies. The next decade holds certainly great excitement and challenges in the development of high efficient microbial processes using ISPR technologies for the industrial manufacture of terpenoids. Biotechnology has never held such a key role in the industrial production of terpenoids like today.

The authors declare no conflict of interest.

References

[1] Marmulla, R & Harder, J. Microbial monoterpene transformation - a review. *Frontier in microbiology.* 2014 doi: 10.3389/fmicb.2014.00346.

[2] Watts, K.T., Mijts, B.N., & Schmidt-Dannert, C. Current and emerging approaches for natural product biosynthesis in microbial cells. *Advanced Synthesis & Catalysis.* 2005, 347(7-8), 927–940.

[3] Mirata, M.A. Microbial cell factories: towards an industrial production of isoprenoids. *ChemCatChem.* 2014, 6, 955-957.

[4] Krings, U. & Berger, R.G. Terpene Bioconversion - How Does its Future Look? *Natural Product Communication.* 2010, 5(9), 1507-1522.

[5] Schwab, W., Fuchs., C., & Huang, F.-C. Transformation of terpenes into fine chemicals. *European Journal of Lipid Science and Technology*, 2013, 115, 3-8.

[6] Kampranisa, S.C., & Makrisb, A.M. Developing a yeast cell factory for the production of terpenoids. *Computational and structural biotechnology journal.* 2012, 3, e201210006.

[7] Brück, T., Kourist, R., & Loll, B. Production of Macrocyclic Sesqui- and Diterpenes in Heterologous Microbial Hosts: A Systems Approach to Harness Nature's Molecular Diversity. *ChemCatChem.* 2014, 6, 1142-1165.

[8] Dafoe, J.T., & Daugulis, A.J. In situ product removal in fermentation systems: improved process performance and rational extractant selection. *Biotechnology Letters.* 2014, 36, 443-460.

[9] Fiedurek, J,, Trytek, M., & Skowronek, M. Strategies for improving the efficiency of bioprocesses involving toxic compounds. *Current Organic Chemistry.* 2012, 16, 2946-2960.

[10] Immethun, C., Hoynes-O'Connor, A., Balassy, A., & Moon, T. Microbial Production of Isoprenoids Enabled by Synthetic Biology. *Frontiers in Microbiology*. 2013, 4(75), 1-8.

[11] Fichian, I., Larroche, C., & Gros, J.B. Water solubility, vapour pressure, and activity coefficients of terpenes and terpenoids. *Journal of Chemical Engineering Data.* 1999, 44, 56-62.

[12] Leng, C., Kish, J., Kelley, J., Mach, M., Hiltner, J., Zhang, Y., & Liu, Y. Temperature-dependent Henry's law constants of atmospheric organics of biogenic origin. *The Journal of Physical Chemistry A.* 2013, 117, 10359-10367.

[13] Mars, A., Gorissen, J., van den Beld, I., & Eggink, G. Bioconversion of limonene to increased concentrations of perillic acid by *Pseudomonas putida* GS1 in a fed-batch reactor. Applied Microbiology and Biotechnology. 2001, 56(1-2), 101-107.

[14] Newman, J., Marshall, J., Chang, M., Nowrozzi, F., Paradise, E., Pitera, D., Newman, K.L., & Keasling, J.D. High-level production of amorpha-

4,11-diene in a two-phase partitioning bioreactor of metabolically engine-ered *Escherichia coli*. Biotechnology and Bioengineering. 2006, 95(4), 684-691.

[15] Wriessnegger, T., Augustin, P., Engleder, M., Leitner, E., Müller, M., Kaluzna, I., Schürmann, M., Mink D., Zellnig, G., Schwab, H., & Pichler, H. Production of the sequiterpene (+)-nootkatone by metabolic engineering of *Pichia pastoris*. *Metabolic Engineering*. 2014, 24, 18-29.

[16] Cal, K. Aqueous solubility of monoterpenes at 293 K and relatioship with calculated log P vaule. *Yakugazu Zasshi*. 2006, 126(4), 307-309.

[17] Schrader, J. Microbial flavour production. In Berger, R.G., editor. *Flavours and fragrances*. Berlin, Heidelberg: Springer, 2007, 509-573.

[18] Brennan, T., Turner, C., Krömer, J., & Nielsen, L. Alleviating monoterpene toxicity using a two-phase extractive fermentation for the bioproduction of jet fuel mixtures in *Saccharomyces cerevisiae*. *Biotechnology and Bioengineering*. 2012, 109(10), 2513-2522.

[19] Sikkema, J., de Bont, J.A. & Poolman, B. Interactions of cyclic hydrocarbons with biological membranes. *The Joural of Biological Chemistry*. 1994, 269(11), 8022-8028.

[20] Mirata, M.A., Wüst, M., Mosandl, A., & Schrader, J. Fungal biotransformation of (+/-)-linalool. *Journal of Agricultural and Food Chemistry*. 2008, 56, 3287-3296.

[21] Mirata, M.A., Heerd, D. & Schrader, J. Integrated bioprocess for the oxidation of limonene to perillic acid with *Pseudomonas putida* DSM 12264. *Process Biochemistry*. 2009, 44, 764-771.

[22] Peralta-Yahya, P., Oullet, M., Chan, R., Mukhopadhyay, A., Keasling, J., & Lee, T. Identification and microbial production of a terpene-based advanced biofuel. *Nature Communication*. 2011, doi:10.1038/ncomms1494.

[23] Buijs, N.A., Siewers, V. & Nielsen, J. Advanced biofuel production by the yeast *Saccharomyces cerevisiae*. *Current Opinion in Chemical Biology*. 2013, 17(3), 480-488.

[24] Stark, D., & von Stockar, U. In situ product removal in whole cell biotechnology during the last twenty years. Advances in Biochemical Engineering and Biotechnology. 2008, 80, 149-175.

[25] Van Hecke, W., Kaur, G. & De Wever, H. Advances in in-situ product recovery (ISPR) in whole cell biotechnology during the last decade. *Biotechnology Advances*. 2014, 32, 1245-1255.

[26] Freeman, A., Woodley, J.M. & Lilly, M.D. In situ product removal as a tool for bioprocessing. *Bio/Technology*. 11, 1007-1012.

[27] Lye, G.J. & Woodley, J.M. Application of in situ product-removal technoques to biocatalytic processes. *Trends in Biotechnology*. 1999, 17, 395-402.

[28] Buijs, N., Siewers, V., & Nielsen, J. In situ product recovery (ISPR) by crystallisation: basic principles, design, and potential applications in whole-cell biocatalysis. *Applied Microbiology and Biotechnology*. 2006, 71, 1-12.

[29] Morrish, J.L., Brennan E.T., Dry H.C., & Daugulis, A.J. Enhanced bioproduction of carvone in a two-liquid-phase partinioning bioreactor with a highly hydrophobic biocatalyst. *Biotechnology and Bioengineering*. 2008, 1(4), 768-75.

[30] Schewe, H., Holtmann, D., & Schrader, J. P450(BM-3)-catalyzed whole-cell biotransformation of alpha-pinene with recombinant *Escherichia coli* in an aqueous-organic two-phase system. *Applied Microbiology and Biotechnology*. 2009, 83(5), 849-857.

[31] Cornelissen, S., Jusling, K., Volmer, J., Reichert, O., Schmid, A., & Bühler, B. Whole-cell-based CYP153A6-catalyzed (*S*)-limonene hydroxylation efficiency depends on host background and profits from monoteprene uptake via AlkL. *Biotechnology and Bioengineering*. 2012, 110(5), 1282-1292.

[32] Tsuruta, H., Paddon, C., Eng, D., Lenihan, J., Horning, T., Anthony, L., Rika Regentin, R., Keasling, J.D., Renninger, N.S. & Newman, J.D. High-level production of amorpha-4,11-diene, a precursor of the antimalarial agent, artemisinin, in *Escherichia coli*. *PLoS one*. 2009, 4(2), e4489.

[33] Paddon, C.J., Westfall, P.J., Pitera, D.J., Benjamin, K., Fisher, K., McPhee, D., Leavell, M.D., Tai, A., Main, A., Eng, D., Polichuk, D.R., Teoh, K.H., Reed, D.W., Treynor, T. Lenihan, J. Jiang, H. Fleck, M., Bajad, S., Dang, G., Dengrove, D., Diola, D., Dorin, G., Ellens, K.W., Fickes, S., Galazzo, J., Gaucher, S.P., Geistlinger, T., Henry, R., Hepp, M., Horning, T., Iqbal, T.,Kizer, L., Lieu, B., Melis, D., Moss, N., Regentin, R.,Secrest, S., Tsuruta, H., Vazquez, R., Westblade, L.F., Xu, L., Yu, M., Zhang, Y., Zhao, L., Lievense, J., Covello, P.S., Keasling, J.D., Reiling, K.K., Renninger, N.S.,& Newman, J.D. High level semi-synthetic production of the potent antimalarial artemisinin. *Nature*. 2013, 496, 528–532.

[34] Schalk, M., Pastore, L., Mirata, M.A., Khim, S., Schouwey, M.,Deguerry, F., Pineda, V., Rocci, L., & Daviet, L. Toward a

biosynthetic route to sclareol & amber odorants. Journal of The American Chemical Society. 2012, 134, 18900-18903.

[35] Willrodt, C., David, C., Cornelissen, S., Bühler, B., Julsing, M., & Schmid, A. Engineering the productivity of recombinant *Escherichia coli* for limonene formation from glycerol in minimal media. Biotechnology Journal. 2014, 9, 1000-1012.

[36] Phillips, T., Chase, M., Wagner, S., Renzi, C., Powell, M., DeAngelo, J., & Michels, P. Use of in situ solid-phase adsorption in microbial natural product fermentation development. *Journal of Industrial Microbiology and Biotechnology*. 2013, 40, 411-425.

[37] Nielsen, D.R. & Prather, K.J. In situ recovery of n-butanol using polymeric resins. *Biotechnology and Bioengineering*. 2008, 102(3), 811-821.

[38] Schrader, J. & Mirata, M.A. Improved bioproduction of perillic acid. *BIOforum Europe*. 2009, 5, 28-29.

[39] Bormann, S., Etschmann, M., Mirata, M., & Schrader, J. Integrated bioprocess for the stereospecific production of linalool oxides from linalool with *Corynespora cassiicola* DSM 62475. Journal of Industrial Biotechnology. 2012, 39(12), 1761-1769.

[40] Alonso-Guiterez, J., Chan, R., Batth, T. A., Kesaling, J., Petzold, C., & Lee, T. Metabolic engineering of *Escherichia coli* for limonene and perillyl alcohol production. *Metabolic Engineering*. 2013, 33–41.

[41] Chen, C., Wang, L., Xiao, G., Liu, Y., Xiao, Z., Deng, Q., & Yao, P. Continuous acetone–butanol–ethanol (ABE) fermentation and gas production under slight pressure in a membrane bioreactor. *Bioressouce Technology*. 2014, 163, 6-11.

[42] Straathof, AJJ. Transformation of biomass into commodity chemicals using enzymes or cells. *Chemical Reviews*. 2014, 114(3), 1871–1908.

[43] Scalcinati, G., Partow, S., Siewers, V., Schalk, M., Daviet, L., & Nielsen, J. Combined metabolic engineering of precursor and co-factor supply to increase α-santalene production by *Saccharomyces cerevisiae*. *Microbial Cell Factory*. 2012, 11(117), 1-16.

In: Terpenoids and Squalene
Editor: Alanna R. Bates

ISBN: 978-1-63463-656-8

Chapter 2

BIOLOGICALLY ACTIVE TRITERPENOIDS USABLE AS PRODRUGS

Milan Urban[1,2]***, Miroslav Kvasnica***[3]***,***
Niall J. Dickinson[1] ***and Jan Sarek,***[1,4*]
[1]Department of Organic Chemistry, Faculty of Science,
Palacky University Olomouc, Olomouc, Czech Republic.
[2]Institute of Molecular and Translational Medicine, Faculty of Medicine
and Dentistry, Palacky University Olomouc, Olomouc, Czech Republic.
[3]Laboratory of Growth Regulators, Palacky University and Institute of
Experimental Botany AS CR, Olomouc, Czech Republic
[3]Betulinines Chemical Group, Stribrna Skalice, Czech Republic

ABSTRACT

Triterpenoids are a large group of natural compounds, many of which have interesting biological activities. Despite this, only a few derivatives are currently used as therapeutics. The main reason is that the active triterpenes usually have unfavorable pharmacological and physicochemical properties such as solubility and bioavailability. Therefore, many compounds with excellent activity in *in vitro* tests usually fail during clinical trial. Many research groups have tried to synthesize a variety of prodrugs in order to improve pharmacological properties of the active compound while retaining the desired biological

[*] E-mail address: jan.sarek@gmail.com, Fax: +420 226 015 452

activity. The most commonly chosen functional groups in triterpenes to be modified as prodrugs are hydroxyl and carboxyl. In this chapter, we are going to summarize all of the research that has been done in the area of triterpenoid prodrugs with the main focus on the modifications of hydroxyls and carboxyls. Commonly used types of prodrug groups (such as simple alkyl esters, other esters, ethers, amides etc.) will be shown as well as unusual prodrugs with more interesting biological and pharmacological properties (glycosides, hemi esters etc.). Their preparation will be shown as well as their chemical, biological, and pharmacological properties. In addition to this, we will compare each type of prodrug with the other prodrug types and will discuss their limits of use. Failures will be shown as well as successful compounds. In concluding, we will show prodrugs with the best properties as well as the current trends in the development of the new triterpenoid prodrugs including the most recent results from our research group.

1. Introduction

Triterpenes are natural compounds that are particularly prevalent in nature, they can especially be found in plants, marine organisms, fungi, molds bacteria etc. Many hundreds of triterpenes are isolated from natural sources every year and Conoly & Hill publish reviews about these new compounds every year, examples from 2012-2013 are in [1-3]. Triterpenes have a variety of biological activities [4]. They are often cytotoxic [5], antiviral [6], antimicrobial [7], antifungal [8], antileishmanial [9], antiplasmodial [10], antiinflammatory [11], antiulcer [12], anticariogenic [13], hepatoprotective [14], cardioprotective [15], analgesic [16] and if someone looks for natural compounds associated with an important biological activity, there will certainly be triterpenes among them. The toxicity of triterpenes is usually low (the therapeutic index is high) which is important for their potential use as therapeutics.

Despite triterpenes having a wide range of biological activities accompanied with low toxicity, the mother compounds usually have two main drawbacks. Firstly; the compounds are not active in low enough concentrations. The IC_{50} usually drops to low micromolar ranges in the most active triterpenes but this is often not enough for them to be used therapeutically because there often are more active compounds in use. This is one of the most important reasons for so many research groups (including us) to be investigating the synthesis of semisynthetic triterpenes in order to increase the activity while retaining low toxicity. Based on some basic and rather intuitive structure-activity relationship (SAR) analysis, these researchers usually design new structures, synthesize them and reevaluate the IC_{50}. Most of the time, they get derivatives slightly more active than their starting compounds but only on rare occasions are the improvements significant. From this point of view, it is a little unfortunate that triterpenes have a lot of biological targets and a lot of activities, because this makes it almost impossible to quantify the SARs and to create and appropriate *in-silico* models to predict the activities of the designed structures accurately. In addition, small changes in the structure of triterpenes can either enhance or completely annihilate their activity; sometimes one activity completely disappears while another emerges. This makes the predictions of the activity of yet unknown compounds difficult.

The second drawback is that triterpenes usually have inappropriate pharmacological properties, most importantly; their solubility in water based media is low. This is one of the main reasons why compounds that have high *in vitro* activities often fail during the *in vivo* experiments and why their administration poses many difficulties.

The first approach to overcome this problem is to prepare conjugates. For example, it has been patented [17], that sodium salts of triterpenoid acids are soluble in water in the presence of 2-hydroxypropyl-γ-cyclodextrine. During our research, however, we found that such conjugates have a lower IC_{50}, probably due to the fact that the complex is either too stable and the compound is not released in the target tissue or the complex is too weak and the triterpene dissociates easily in plasma and then re-associates with plasma proteins. It is in agreement with the literature, for example [18], that the stability of the complexes depends on the structure and modifications of both triterpenes and cyclodextrines.

The second approach is to synthesize appropriate prodrugs. In the following text, the most commonly used prodrugs made from triterpenes, their synthesis, properties and their influence on the biological activities will be discussed. By definition, a prodrug is a compound that is administered in its inactive (or less active) form that has appropriate pharmacological properties and thus is more accessible to the target tissue. After it reaches its target tissue, the prodrug undergoes a metabolic transformation which converts it to the active form. In triterpenes, not many studies of their metabolism have been published. Therefore for the purposes of this chapter, any scaffold attached to a terpenic hydroxyl or carboxyl that likely undergoes metabolic transformation in the target tissues will be included.

2. PRODRUGS OF TRITERPENES

As mentioned before, prodrugs include both triterpenic carboxylic acid derivatives (esters and amides) and/or alcohol derivatives (ethers, esters, hemiesters and glycosides).

2.1. Derivatives of Triterpenic Carboxylic Acids

2.1.1. Alkyl and Benzyl Esters

Esters are the most commonly used groups to protect carboxylic acids, and since they often undergo hydrolysis by nonspecific esterases in cells, they are often used as prodrugs. Esters are less polar than the starting acids, which makes them less soluble in water based media; and conversely, may enhance their transport across biological membranes. In triterpenes, esters are commonly studied because it is simple to synthesize them by alkylation of the starting acid. Diazoalkane, haloalkane and other alkylating agents may be used to prepare esters. The most commonly used alkylesters are methyl-, ethyl-, and benzyl-ester; Figure 1. Their influence upon biological activity varies with the structure of the triterpene and it usually corresponds with their stability: the more stable the ester is, the lower activity it has. It seems that a free carboxyl group is commonly an essential part of the pharmacophore, and if we bind them in an ester that may not be hydrolyzed within the target tissue, the activity is lost.

In lupane or lup-20(29)-ene acids such as betulinic acid (1), the presence of an alkylester at 28-COOH completely diminishes the activity. Due to steric hindrance (neopentyl type of acid), this ester is extremely stable and to deprotect it chemically, 10% KOH in refluxing ethylene glycol is required. Obviously, no enzyme is capable of doing this. Cytotoxic activity of 1 is usually in the range of IC_{50} 15-20 μM on multiple cancer cell lines while the IC_{50} of 2-4 is in high micromolar ranges over 100 μM [19, 20]. On the other hand, there are many examples of triterpenic methyl esters that are biologically active. We described alkyl 21,22-oxolup-18-ene-28-oates such as 5 that are highly cytotoxic with an IC_{50} of 1-5 μM on many cancer lines [21]. This type of ester, however, is much easier to deprotect. The presence of the 18-double bond and the 21,22-ketones "flattens" the cycle and makes the carboxyl group electron deficient and thus, easily cleavable. In order to chemically remove the ester group, water with a presence of hydroxide in refluxing alcohol is sufficient.

Other examples of active alkyl esters include derivatives of oleanane such as bardoxolone methyl 6 that can suppress oxidative stress and inflammation and is an inducer of Nrf2 pathway [22]; and maslinic acid (7) and its esters such as 8-24 prepared to study [23] the influence of the ester group on the cytotoxic activity of maslinic acid derivatives. Lit. [23] concluded the cytotoxicity was increased by the presence of a short hydrophobic ester (such as methyl) while diminished by long chain esters. When the alkylester is prepared from a different carboxyl than 28-COOH, the activity is usually similar to the starting acid. For example, there was no influence on the anti-arthritic activity between glycyrrhetic acid and its methyl ester [24]. In conclusion, using triterpenic alkyl esters as prodrugs is sometimes appropriate. The advantage of this approach is their easy and cheap preparation.

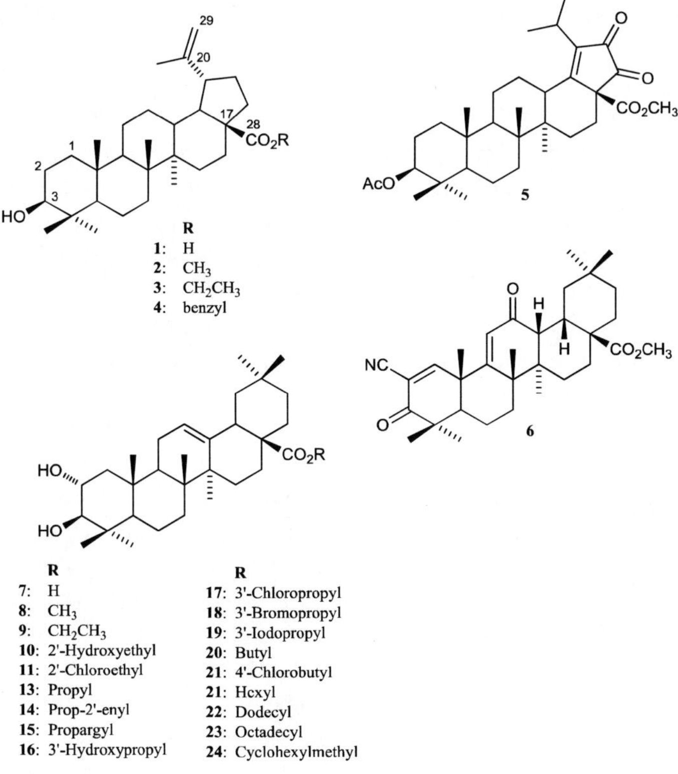

Figure 1. Triterpenic alkyl esters.

Ethyl esters are not as common as methyl esters, though they have been reported to provide better biological activity than methyl esters in some cases [25]. Benzyl esters are usually used as protecting groups for the synthetic modifications of triterpenic acids. They are simple to synthesize by alkylation reactions and they are simple to deprotect by hydrogenolysis. Even in compounds with double bond 20(29), the use of benzyl ester is possible because the double bond remains unsaturated under atmospheric pressure of hydrogen gas and with the use of Pd/C catalyst. They have a lot of advantages in the chemical synthesis, however, they are almost always inactive, and therefore not suitable as prodrugs. Halogenated and hydroxylated alkyl esters such as 10, 11, 16-19, 21 had activity in similar ranges as the starting acid [23]. We may expect that other pharmacological properties such as solubility may be slightly enhanced and therefore they seem to be interesting prodrugs. Also, it is likely that they are reactive enough to undergo enzymatic transformations, however, no such study was reported to support such hypothesis.

2.1.2. Other Ester Types

To overcome the extreme stability of methylester of lupane acids, an effort has been made by our group to prepare acyloxymethylesters; Figure 2. These ester types are well known from penicillin derivatives [26] as well as from nucleoside prodrugs [27]. It is known, that pivaloyloxymethylesters (Pom) and acetoxymethylesters (Acm) are cleaved by nonspecific esterases in the target cells [28]. We prepared [20, 29] a set of Pom and Acm esters such as 25-28 from active derivatives of betulinic and dihydrobetulinic acid by alkylation of them with Pom-Cl or Acm-Br. We found Acm esters to be as active as the free acids while Pom esters inactive, obviously because they are sterically hindered from both sides. In more modified and less stericaly hindered triterpenes, both Pom and Acm esters tend to be active. Acm esters, thus, were found to be suitable prodrugs. They did not decrease the activity while they improved the other properties of the molecules. This type of a prodrug, however, did not solve the problems with low solubility in water based media. Acm esters were also used in [30] to enhance the activities of triterpenoid derivatives.

Patent [30] also describes preparation and use of esters of betulinic acid (1) with cholesterol 29 in *in vivo* experiments. While cholesterol molecule improves the bioavailability of the triterpene, it also decreases solubility in water. By the use of polyethylene glycol substituent at C-3, for example 30, a compound easy soluble in water was obtained.

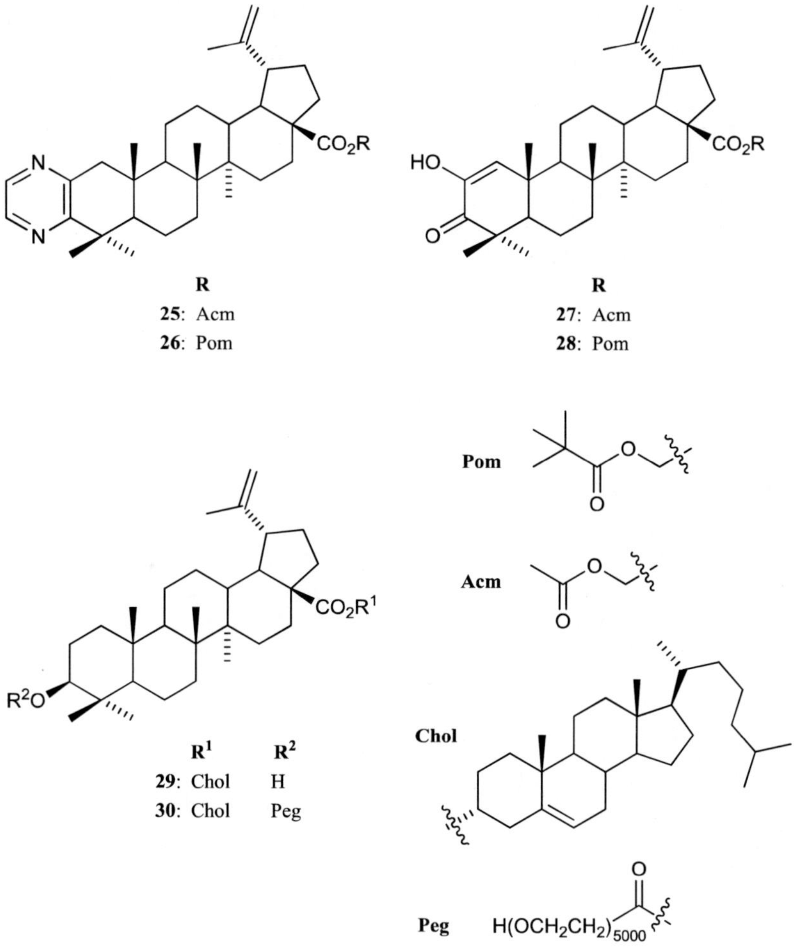

R
25: Acm
26: Pom

R
27: Acm
28: Pom

Pom

Acm

R^1	R^2
29: Chol	H
30: Chol	Peg

Chol

Peg $H(OCH_2CH_2)_{5000}$

Figure 2. Enzymatically cleavable triterpenic esters.

Other types of esters that have been prepared to enhance the solubility of active triterpenic acids are amino esters such as 33-36 in [31] that contain amine or quaternary ammonium salts. In [31], such derivatives were prepared from betulinic, dihydrobetulinic, platanic, oleanolic, ursolic, and 3β-acetoxy-21-oxolup-18-en-28-oic and in some cases, the cytotoxic activity was lower than the activity of the corresponding acid. In other cases, such as derivatives of dihydrobetulinic acid (31), the resulting esters had the activity on multiple cell lines similar or better than the starting material; Figure 3. Aminoesters [32] and hydroxyaminoesters [33, 34] have also been prepared from betulinic acid (1). In [32], a set of compounds were prepared containing a heterocyclic amine such as 37-39 with the variety in the length of the linker between 2-4 atoms and the amine at the terminus.

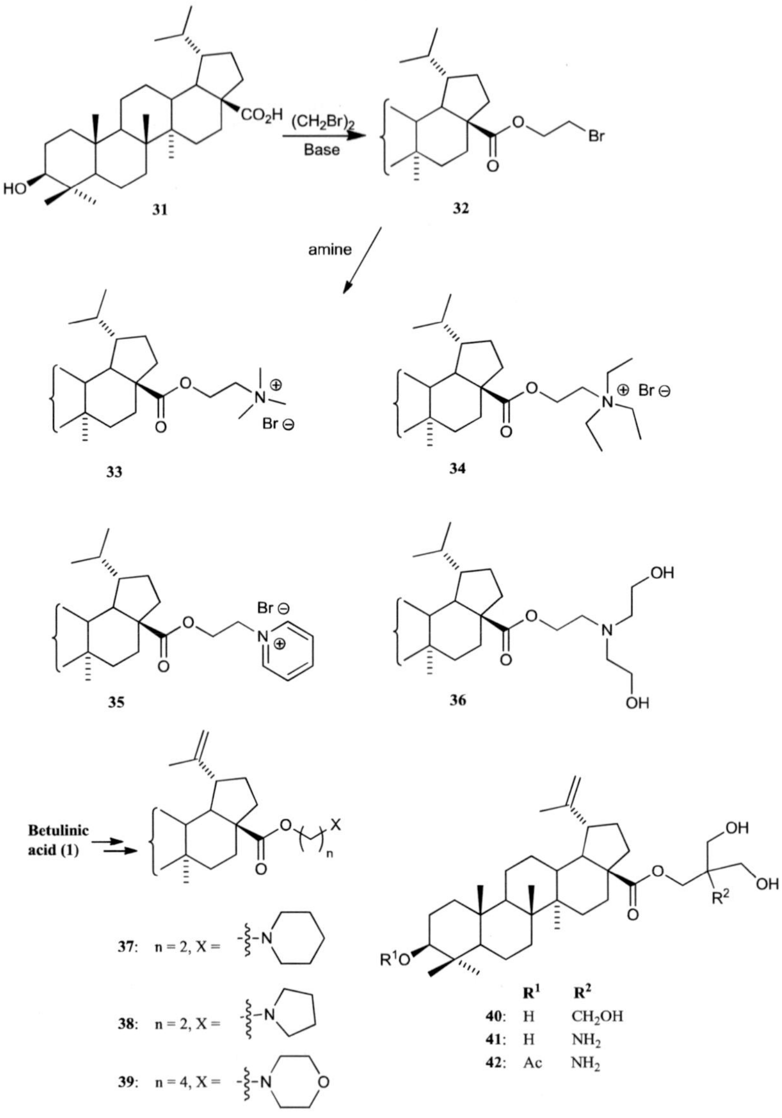

	R^1	R^2
40:	H	CH_2OH
41:	H	NH_2
42:	Ac	NH_2

37: n = 2, X = piperidin-1-yl

38: n = 2, X = pyrrolidin-1-yl

39: n = 4, X = morpholin-4-yl

Figure 3. Triterpenic esters containing quaternary ammonium salts or amine.

Some compounds were cytotoxic in the same level of IC_{50} as the starting betulinic acid (1), however, compound 37 had an IC_{50} of 2.3-4.6 μM on five cancer cell lines, which is significantly better than the IC_{50} of betulinic acid (1) (around 18 μM for those lines). In [33], even more polar compounds were synthesized containing several free hydroxyl or amine moieties of general formula 40. The authors found that compound 41 was the most active among the studied derivatives causing apoptosis in colon cancer cell line HT-29. In [34], derivative 42 (NVX-207) was selected from a larger set of derivatives, and studied in multiple biological tests. It had an IC_{50} of 3.5 μM on multiple human and canine cancer cell lines. Paper [34] concludes that compound 42 induced apoptosis, they explained the mechanism of action and found the compound to have excellent results in clinical responses including a complete remission in treated canine.

Sugar esters are another type of prodrug that have been used in triterpenes, see figure 4. However, this type of ester has only been used rarely [35]. It is supposed that the presence of a sugar moiety at 28-COOH lowered the activity as in case of 43.

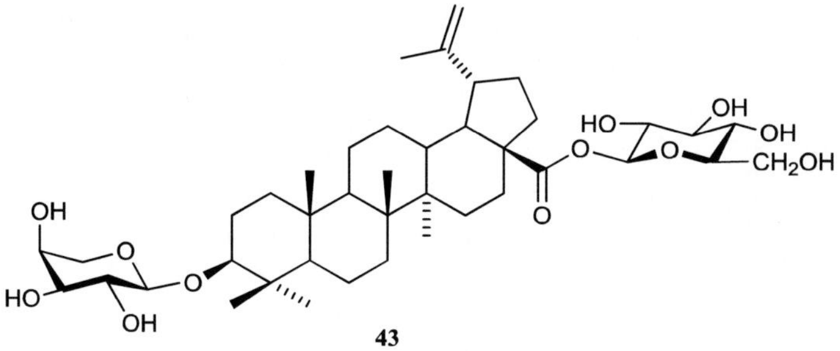

OH
O
HO
OH
O
HO
O
O
OH
OH
CH_2OH
O
43

Figure 4. Example of triterpenic monosaccharide ester.

2.1.3. Amides

Amides are very commonly prepared from triterpenoid acids and there are a large variety of amide structures. Amides often enhance the activity of the original acid. A set of modified amides of betulinic acid 44-52 were synthesized for example in the lit. [36] within a large study of triterpenoid derivatives and their effects on G-protein coupled receptor TGR5 activity. It was found that all of the amides had lower activity than the parent acid. Another large study of amides of triterpenic acids are in [37, 38]. Lit. [37] describes a series of betulinic acid amides such as 53-60 and their anti-HIV activity; finding the lowest IC_{50} in compounds 53 and 57 (3.0 μM and 0.29

μM). In [38], a series of triterpenoid amides (61 and 62) were studied for their inhibition activity of proteasome.

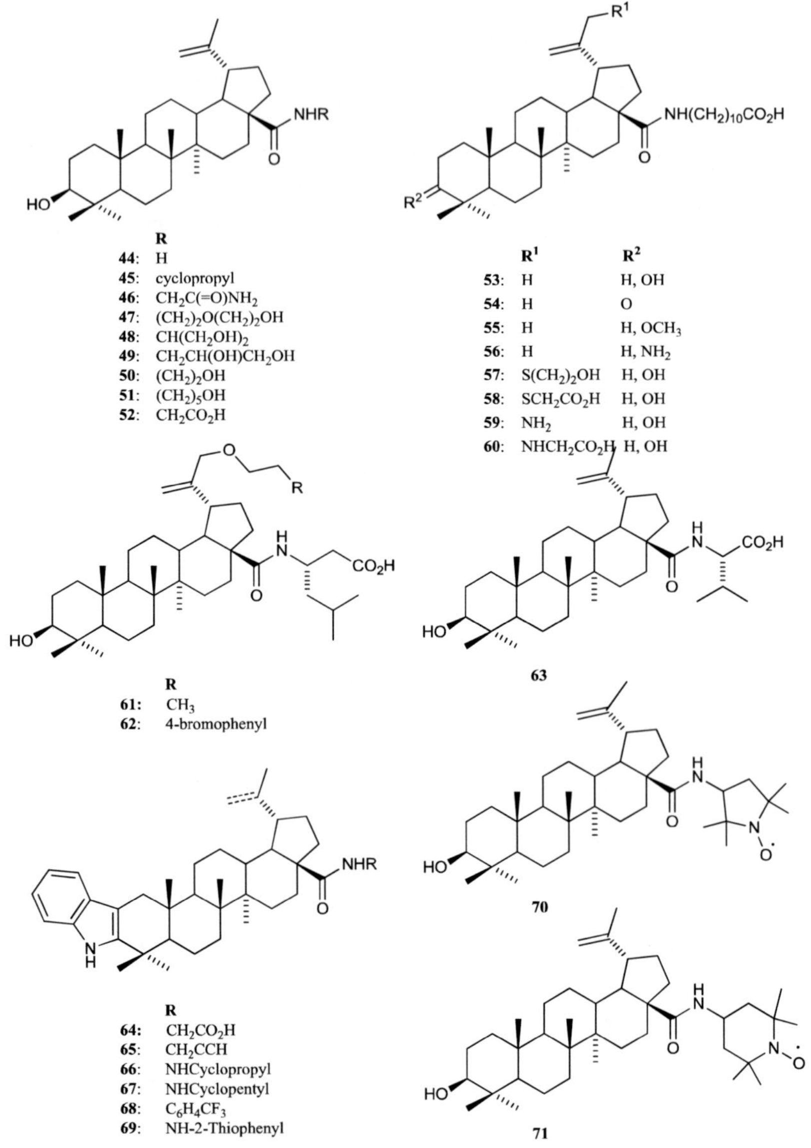

Figure 5. Triterpenic amides.

The amide group was combined with substitution on C-3 and C-30 and the most active compounds were amide derivatives 61. Lit. [39] describes a set of conjugates and cytotoxic activity of betulinic acid (1) with natural amino acids, with the most active to be valine derivative 63. In the study [39], the presence of the amino acid amide increased the solubility of the triterpene and also increased the activity. There are many literature examples, where a biological activity of semisynthetic triterpenes was enhanced by transforming the 28-COOH to a variety of amides such as compounds 64-69 in the lit. [40]. Interesting amides with more modifications around the terpenic skeleton may be found in [41-43]. In the lit [41], there are highly cytotoxic amides containing fragments with nitroxyl radicals 70 and 71, in [42], the compounds had no activity and in [43], the amides had antiherpes and anti-HIV activity with EC_{50} in the low micromolar ranges. For examples of triterpenic amides, see Figure 5.

2.2. Derivatives of Triterpenic Alcohols

2.2.1. Alkyl And Aryl Ethers and Silylethers

Ethers may be synthesized by alkylation of triterpenic alcohols with haloalkanes under strongly basic (e.g. sodium hydride) conditions. Ethers in general are not suitable prodrugs, because they decrease water solubility and the biological activities of ethers are always low. They do not undergo enzymatic cleavage and the chemical deprotection is only possible under very harsh conditions (strong protic acids or Lewis acids), which often cause skeletal rearrangements. The main use of silyl ethers of triterpenes is as protecting groups for chemical transformations of triterpenes but not as prodrugs, as described in the literature [37].

2.2.2. Esters Derived from Alkyl Acids

Esters are again the most commonly used protection of triterpenic alcohols. Esters with monocarboxylic acids (fatty acids) are the most commonly used. They are easily accessible by a reaction of triterpenic alcohol with anhydrides or acyl halides in the presence of base. Acetates and propionates are often used; they usually decrease the solubility of the prodrugs in water and decrease the biological activities. They probably undergo enzymatic hydrolysis which may be too slow to keep the activity of unprotected alcohols. During our work, we found, that most of the time, acetates decrease the biological activity slightly while chloroacetates have similar cytotoxicity to the parent alcohols [21, 25]. There is an exception to that rule, acetates of some triterpenoid saponines (e.g. 100) are more active than the deacetylated derivative (101) [17]. Other common types are esters with fatty acids with a long carbohydrate chain and their preparation was published in [44, 45]. Compounds of the formulas 72 – 74 had significant antibacterial activities against Gram negative bacteria *Pseudomonas syringae* (ATCC #13457) and fairly good activity against Gram positive bacteria, *Bacillus sphaericus* (ATCC #14577) and *Bacillus subtilis* (ATCC #6051) in [44]. Similar derivatives of ursolic and oleanolic acid were proven to have antifeedant activity against an agricultural pest, tobacco cutworm larvae (*Spodoptera litura* F) in [45].

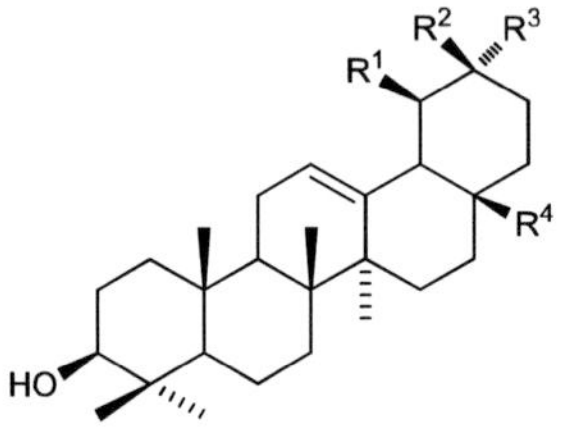

72: $R^1 = R^2 = R^4 = CH_3$, $R^3 = H$
73: $R^1 = H$, $R^2 = R^3 = R^4 = CH_3$
75: $R^1 = R^3 = CH_3$, $R^2 = H$, $R^4 = CO_2H$

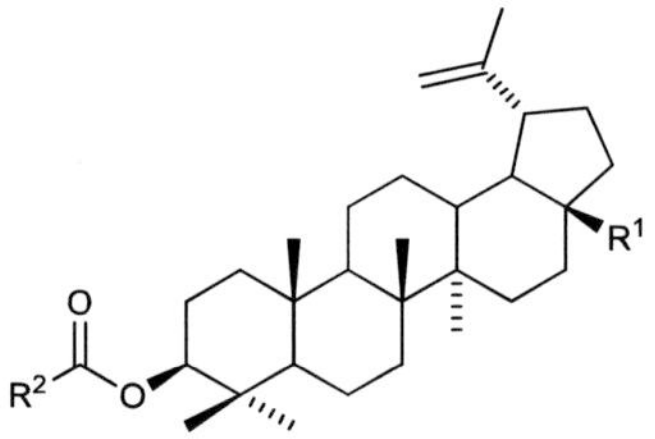

77: $R^1 = CO_2H$ R^2 = 3-chlorophenyl
78: $R^1 = CH_3$ R^2 = 4-chlorophenyl
79: $R^1 = CH_3$ R^2 = 4-nitrophenyl
80: $R^1 = CO_2H$ R^2 = 3-pyridyl

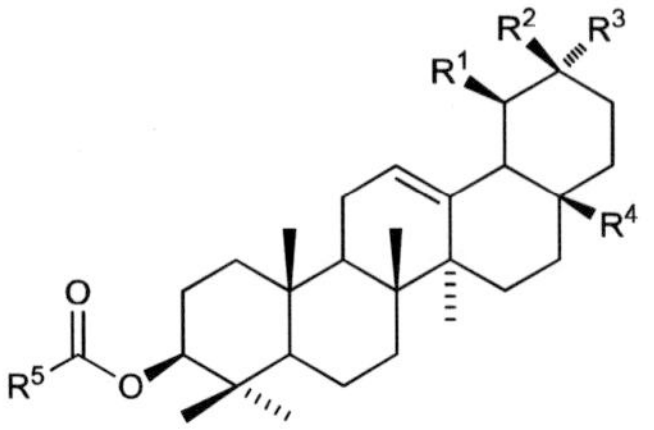

74: $R^1 = R^3 = CH_3$, $R^2 = H$, $R^4 = CO_2H$, $R^5 = CH_3(CH_2)_{10}$
76: $R^1 = H$, $R^2 = R^3 = CH_3$, $R^4 = CO_2H$, R^5 = 3-chlorophenyl

81. $R_1 = CH_3$ $R^2 =$ 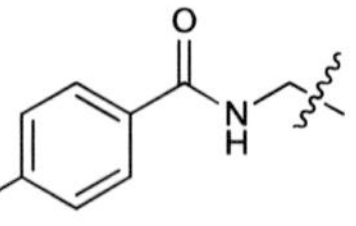
82. $R_1 = CH_3$ $R^2 =$ 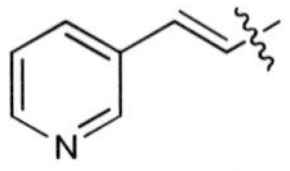
83. $R_1 = CH_3$ $R^2 =$ 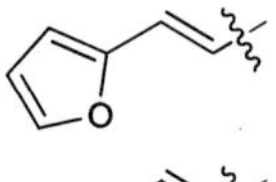
84. $R_1 = CH_3$ $R^2 =$ 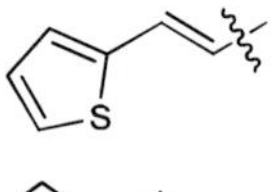
85. $R_1 = CH_3$ $R^2 =$

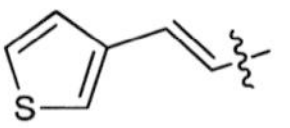

Figure 6. Esters of triterpenic alcohols.

2.2.3. Esters Derived from Aryl Acids

Esters with aromatic carboxylic acids are also very common prodrugs of triterpenes. In [46], chlorobenzoates such as 76-77 were synthesized from betulinic, ursolic, and oleanolic acids and studied for their antioxidant level, *Artemia salina* lethality and antimicrobial activity. A large set of aromatic acid esters were prepared in [47], similarly to [46], they published lupeol chlorobenzoates and nitrobenzoates e.g., 78 and 79 but also esters of acids containing heterocycles such as pyridine, thiophene, and furan, see formulas 80 – 85. The derivative 80 decreased the level of blood glucose by 18.2% and 25.0% at 5 h and 24 h, respectively and also lowered 40% (P <0.001) in triglycerides, 30% (P <0.05) in glycerol, 24% (P <0.05) in cholesterol levels and also improved the HDL–cholesterol level by 5% in dyslipidemic hamster model at the used dose of 50 mg/kg b.wt [47]. Another variety of aromatic acid esters are cinnamates and their analogs. A large study containing various cinnamate esters of betulinic acid, ursolic acid and oleanolic acid (e. g., 86 and 87) has been published [48]. Several derivatives were found to have significant activity against *Mycobacterium tuberculosis*. A set of esters of triterpenes with nicotinic acid were synthesized in [49], oleanolic, ursolic and glycyrrhetic acids were esterified at their alcohols on C-3, betulin were esterified on both hydroxyls at C-3 and C-28 to form bisnicotinate 88 in figure 7. Betulin 28-methoxycinnamate had high activity against the influenza type A (H1N1) virus with the selectivity index SI > 100 and betulin dinicotinate was active against the papilloma virus (strain HPV-11) with the selectivity index of SI 35.

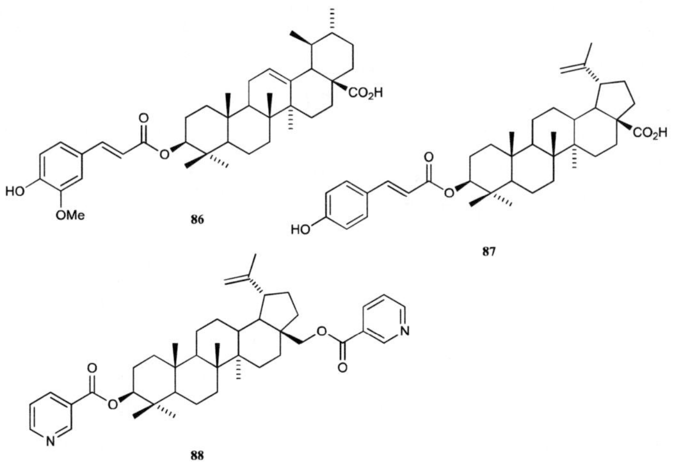

Figure 7. Triterpenic cinnamate and nicotinate esters.

2.2.4. Hemiesters (Esters Derived from Dicarboxylic Acids)

These are very commonly used prodrugs in triterpene chemistry. Such esters allow retaining a polar carboxyl group in the terpene which enhances the solubility in water based media and also allows preparing salts that may be fully soluble in water by adding 2-hydroxypropyl-γ-cyclodextrines and similar solubility enhancers. The most important representative of this group is bevirimat (89), a 3',3'-dimethylhemisuccinate of betulinic acid which is an extremely potent anti-HIV agent. In Bevirimat (89), however, the hemiester is not a prodrug, since the hemiester is the essential part of the pharmacophore and does not undergo any enzymatic transformation in order to release the active compound. The hemiester is active itself. In other compounds, however, the hemiester may be used as a solubility enhancer of an active derivative.

A series of hemiphthalates of betulin and betulinic acid (1) together with their cytotoxic activity on multiple cancer cell lines was described in [50]. The authors studied a combination of various esters at the 28-COOH and at the hemiphthalate acid and their influence on the biological activity, compounds 90-97 are examples of the studied esters but many more combinations were published [50].

Most of the compounds were more active than the starting material with the best results in compound 95 with an IC_{50} of 5-13 μM on several cancer cell lines. Hemisuccinates, hemiglutarates, dimethylhemiglutarates along with the previously mentioned dimethylhemisuccinates were used as well. Another large study on hemiesters is in [51] which contains hemiphthalates, hemisuccinates, acetyl salicylates, and cinnamates of betulin and lupeol and their hepatoprotective activity. In most cases, the biological activity of hemiesters remained in the same range of IC_{50} as for the parent compounds while the pharmacological properties improved. It is not known if such esters undergo any enzymatic transformation but it is likely that some of them may hydrolyze while another possibly may be conjugation with small polar molecules such as amino acids or sugars (Figure 8).

2.2.5. Triterpenoid Saponins

A lot of triterpenes naturally occur as saponins. Saponins are usually soluble in water and if they contain an active aglycone, the molecule has both, favorable IC_{50} and good pharmacological properties. This is one of the main reasons for the synthesis of saponins and deoxyglycosides from active semisynthetic terpenic alcohols. In the publication of Baltina et al. [52], a set of 3-*O*-galactopyranosides of glycyrrhetic acid such as 98 were prepared and their anti-HIV activity studied.

89

	R^1	R^2		R^1	R^2
90:	CO_2H	H	**94**:	CH_2OPhth	H
91:	CO_2Me	H	**95**:	CO_2H	Me
92:	CO_2Et	H	**96**:	CO_2H	Et
93:	CO_2Bn	H	**97**:	CO_2Me	Me

Figure 8. Triterpenic hemiesters.

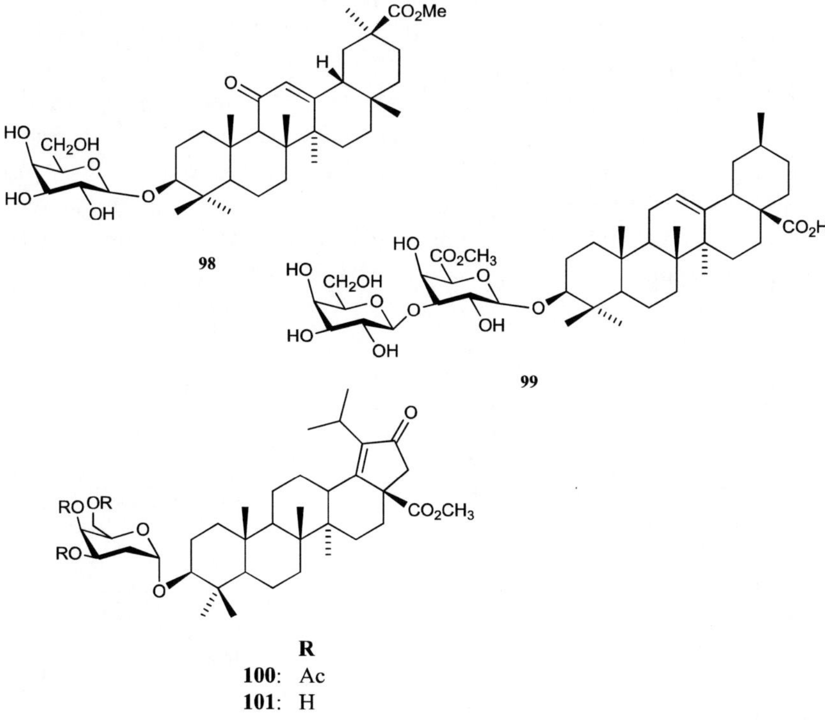

Figure 9. Triterpenic glycosides.

These analogs of glycyrrhetic acid were about 2.5-times more potent as HIV inhibitors than the original compound. Another study [53] of triterpenoid glycosides contains glycosides of oleanane type, compound 99 had the highest cytotoxic activity on MCF-7 cell line.

In our research group, a set of 2-deoxyglycosides of triterpenoid (the best of 2-deoxygalactoside 100) was found to have cytotoxic activity in submicro molar range on multiple tumor cell lines and favorable pharmacological properties with increased solubility than that of the aglycones, therefore it was patented [54].

2.2.6. Labile Esters (as Sulfates)

Sulfates are probably the simplest functional group to enhance solubility. Sulfates are very common in steroid chemistry, therefore it is surprising that there are only few examples of triterpenoid sulfates published in the literature. In [55], synthesis of 3β-*O*-sulfates of oleanolic, echinocystic, and betulinic acid (1) such as 102 was described with the use of complexes of SO_3 with a variety of solvents. However, only the use of DMSO as a reaction solvent afforded reasonable yields.

No biological activities were reported [55]. A more recent paper [56] describes the synthesis of sulfates at 3, and 16 hydroxyls of lupane derivative calenduladiol (103 and 104) and the authors investigated the anticholinesterase activities of their products. However, the most active compound in this set of derivatives was an analog that did not contain any sulfate. The authors used an $Me_3N.SO_3$ complex in DMF.

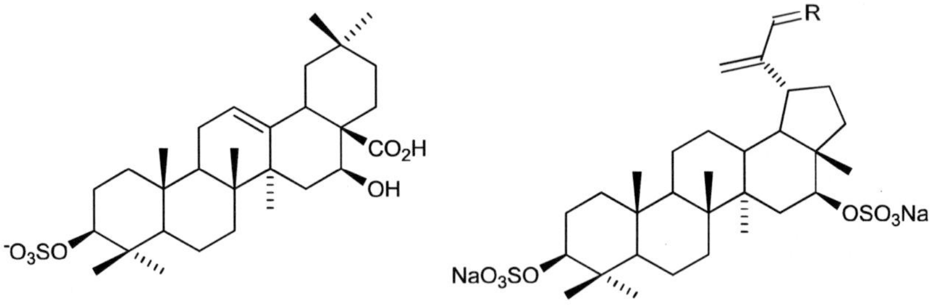

^-O_3SO CO₂H OH

102

NaO_3SO OSO_3Na R

	R
103:	H, H
104:	O

Figure 10. Triterpenic sulfates.

2.3. Combined Derivatives

Triterpenoids almost always have more than one functional group suitable for an attachment of a prodrug function. Many papers were dedicated to such compounds that modify both the 28-COOH (or 28-CH_2OH) and 3-OH positions, in order to improve the pharmacological properties and to further increase the biological activities than when only one group is substituted. In [57], there is a large set of various amides and hemiesters (examples are compounds 105-109) prepared from oleanolic acid which were proven to have an inhibitory activity on dimerization of HIV-1 protease.

The authors investigated the influence of the length of the hemiester chain on the activity, and found that the optimum length is 6-8 carbons. Hemiesters (such as 105) were in the same range of activity as their amide analogs (e.g. 108) while adding an amide or ester with a longer chain to 28-COOH increased the activity significantly. In this paper, it is more likely this increase is as a result of an optimization of the pharmacophore than an effect of a prodrug.

Huang et al. [58] tried to combine two anti-HIV pharmacophores on betulinic acid (1); a 3',3'-dimethylhemisuccinate from bevirimat (89) and the amide from IC9564, and synthesized compound 110. According to the lit. [59] both modifications seem to be having the desired effect - in compound 110, the hemisuccinate possibly prevents the HIV maturation, while the amide hinders the HIV entrance to the cell. The ED_{50} was 2.6 nM, which is a lower dose than both of the original compounds.

A similar study has been published on moronic acid where the best derivative had an ED_{50} of 8.5 nM. It is interesting to observe a similar approach appears to work well on a variety of triterpenoid structures. Cheng et al. [60] studied the influence of various amides at C-28 on the activity of saponin β-hederin 111. They synthesized a set of amide derivatives such as 112-114 and they found compound 113 to be about twice as cytotoxic on several cancer cell lines than the starting 111. This is a nice example of a combination of a glycoside moiety in position 3 with amides in position 28 of the triterpenoid skeleton.

In [61], there is a similar study that combines nicotinates at 28-COOH and glycoside moiety in 18α-glycyrrhetic acid, however, the change in stereochemistry had a positive effect on cytotoxicity and a negative effect on anti-HIV activity.

105: n = 4; R = H or Me
106: n = 1; R = $C(CH_3)_2CO_2H$
107: n = 4; R = CH_3

108: R = OH X = NH
109: R = $NH(CH_2)_5CO_2H$ X = O

110

	R
111:	OH
112:	$NH(CH_2)_2CH_3$
113:	-N(piperazine)NH
114:	-N(piperazine)NCH_3

Figure 11. Examples of combined derivatives.

3. Future Perspectives

This chapter is intended as an insight into the topic of triterpenoid prodrugs. Since not many enzymatic studies have been published, it is not certain which modifications undergo enzymatic changes and which do not. Therefore we included all possible transformations of triterpenoid hydroxyls and carboxyls whether or not they are the “true prodrugs”; and we are simply discussing whether the presence of each prodrug group had any influence on the biological properties and solubility if known. Triterpenes have a lot of biological activities but they often cannot be used in biological tests, especially in *in vivo* tests, because of their unfavorable physicochemical and pharmacological properties. Most commonly, their solubility is too low. Therefore, a lot of hydrophilic triterpenoid derivatives have been prepared from the free alcohols and acids and in many cases, these properties improved significantly. For many active derivatives, however, the appropriate prodrug has not been found yet and more research is being devoted to reach this goal. The significance of triterpenic prodrugs may also be represented by many patents that should protect the best prodrugs already found. One of the largest patents [62] contains a lot of prodrugs, many of them are solubility enhancers such as PEG esters etc. and it protects the use of the prodrugs as cancer and HIV therapeutics. Despite the large numbers of derivatives intended to be used as prodrugs, there are no *in vivo* experiments nor there are other biological studies about the enzymatic stability of those compounds.

ACKNOWLEDGMENTS

This work was supported from the Operational Program Research and Development for Innovations (project CZ.1.05/2.1.00/01.0030) and NPU I projects LO1304, LO1204 and from Ministry of Education, Youth and Sport of the Czech Republic and by the European Social Fund (CZ.1.07/2.3.00/30.0060).

REFERENCES

[1] R. A. Hill and J. D. Connolly, Triterpenoids. *Nat. Prod. Rep.* 28, 1087 (2012).

[2] R. A. Hill and J. D. Connolly, Triterpenoids. *Nat. Prod. Rep.* 29, 780 (2012).

[3] R. A. Hill and J. D. Connolly, Triterpenoids. *Nat. Prod. Rep.* 30, 1028 (2013).

[4] P. Dzubak, M. Hajduch, D. Vydra, A. Hustova, M. Kvasnica, D. Biedermann, L. Markova, M. Urban, and J. Sarek, Pharmacological

activities of natural triterpenoids and their therapeutic implications. *Nat. Prod. Rep.* 23, 394 (2006).

[5] J. Sarek, M. Kvasnica, M. Vlk, and D. Biedermann, Semisynthetic lupane triterpenes with cytotoxic activity (chapter 6). *In Pentacyclic Triterpenes as Promising Agents in Cancer*, ed.: J. A. R. Salvador, Nova Science Publishers, New York, pp 159-189 (2010).

[6] Z. Dang, P. Ho, L. Zhu, K. Qian, K.-H. Lee, L. Huang, and C.-H. Chem, New Betulinic Acid Derivatives for Bevirimat-Resistant Human Immunodeficiency Virus Type-1. *J. Med. Chem.* 56, 2029 (2013).

[7] W.-J. Zuo, H.-F. Dai, J. Chen, H.-Q. Chen, Y.-X. Zhao, W.-L. Mei, and J.-H. Wang, Triterpenes and Triterpenoid Saponins from the Leaves of Ilex kudincha. *Planta Medica* 77, 1835 (2011).

[8] B. Innocente, B. B. Casanova, F. Klein, A. D. Lana, D. Pereira, M. N. Muniz, G. G. Sonnet, A. M. Fuentefria, and S. C. B. Gnoatto, Synthesis of Isosteric Triterpenoid Derivatives and Antifungal Activity. *Chemical Biology & Drug Design* 83, 344 (2014).

[9] R. Haavikko, A. Nasereddin, N. Sacerdoti-Sierra, D. Kopelyanskiy, S. Alakurtti, M. Tikka, C. L. Jaffe, and J. Yli-Kauhaluoma, Heterocycle-fused lupane triterpenoids inhibit *Leishmania donovani* amastigotes. *Med. Chem. Comm.* 5, 445 (2014).

[10] G. Chianese, S. R. Yerbanga, L. Lucantoni, A. Habluetzel, N. Basilico, D. Taramelli, E. Fattorusso, and O. Taglialatela-Scafati, Antiplasmodial Triterpenoids from the Fruits of Neem, Azadirachta indica. *J. Nat. Prod.* 73, 1448 (2010).

[11] Y. Fu, E. Zhou, Z. Wei, D. Liang, W. Wang, T. Wang, M. Guo, N. Zhang, and Z. Yang, Glycyrrhizin inhibits the inflammatory response in mouse mammary epithelial cells and a mouse mastitis model. *The FEBS Journal* 281, 2543 (2014).

[12] S. Yano, M. Harada, K. Watanabe, K. Nakamaru, Y. Hatakeyama, S. Shibata, K. Takahashi, T. Mori, K. Hirabayashi, M. Takeda, and N. Nagata, Antiulcer Activity of Glycyrrhetic Acid Derivatives in Experimental Gastric Lesion Models. *Chem. Pharm. Bull.* 37, 2500 (1989).

[13] N. Joycharat, S. Thammayong, S. Limsuwan, S. Homlaead, S. P. Voravuthikunchai, B. Yingyongnarongkul, S. Dej-adisai, and S. Subhadhirasakul, Antibacterial substances from Albizia myriophylla wood against cariogenic Streptococcus mutans. *Archives of Pharmacal Research* 36, 723 (2013).

[14] T. Morikawa, K. Ninomiya, K. Imura, T. Yamaguchi, Y. Akagi, M. Yoshikawa, T. Hayakawa, and O. Muraoka, Hepatoprotective medicine from traditional Tibetan medicine Potentilla anserina. *Phytochemistry* 102, 169 (2014).

[15] C. Sanchez-Quesada, A. Lopez-Biedma, F. Warleta, M. Campos, G. Beltran, and J. J. Gaforio, Bioactive Properties of the Main Triterpenes Found in Olives, Virgin Olive Oil, and Leaves of Olea europaea. *J. Agric. Food Chem.* 61, 12173 (2013).

[16] M. Gaertner, L. Muller, J. F. Ross, A. R. S. Santos, R. Nieto, J. B. Calixto, R. A. Yunes, F. Delle Monache, and V. Cechinel-Filho, Analgesic Triterpenes from Sebastiania schottiana roots. *Phytomedicine* 6, 41 (1999).

[17] J. Sarek, M. Hajduch, M. Svoboda, K. Novakova, P. Spacilova, T. Kubelka, and D. Biedermann, Method of Preparation of a Soluble Formulation of Water-Insoluble Pentacyclic and Tetracyclic Terpenoids, A Soluble Formulation of a Pentacyclic or Tetracyclic Terpenoid and a Pharmaceutical Composition Containing this Soluble Formulation. US 8653056, Feb. 18, 2014.

[18] B. Claude, P. Morin, M. Lafosse, and P. Andre, Evaluation of apparent formation constants of pentacyclic triterpene acids complexes with derivatized β- and γ-cyclodextrins by reversed phase liquid chromatography. *J. Chromatography A* 1049, 37 (2004).

[19] M. Urban, J. Sarek, J. Klinot, G. Korinkova, and M. Hajduch, Synthesis of A-Secoderivatives of Betulinic Acid with Cytotoxic Activity. *J. Nat. Prod.* 67, 1100 (2004).

[20] M. Urban, J. Sarek, I. Tislerova, P. Dzubak, and M. Hajduch, Influence of esterification and modification of A ring in a group of lupane acids on their cytotoxicity. *Bioorg. Med. Chem.* 13, 5527 (2005).

[21] J. Sarek, J. Klinot, P. Dzubak, E. Klinotova, V. Noskova, V. Krecek, G. Korinkova, J. O. Thomson, A. Janostakova, S. Wang, S. Parsons, P. M. Fisher, N. Z. Zhelev, and M. Hajduch, New Lupane Derived Compounds with Pro-Apoptotic Activity in Cancer Cells: Synthesis and Structure-Activity Relationships. *J. Med. Chem.* 46, 5402 (2003).

[22] L. Fu and G. W. Gribble, Efficient and Scalable Synthesis of Bardoxolone Methyl (CDDO-methyl Ester). *Org. Lett.* 15, 1622 (2013).

[23] B. Siwert E. Pianowski, and R. Csuk, Esters and amides of maslinic acid trigger apoptosis in human tumor cels and alter their mode of actionwith respect to the substitution pattern at C-28. *Eur. J. Med. Chem.* 70, 259 (2013).

[24] M. Alanazi, M. A. Al-Omar, M. M. Abdulla, and A.-G. Amr, Anti-arthritic and immunosuppressive activities of substituted triterpenoidal candidates. International *Journal of Biological Macromolecules* 58, 245 (2013).

[25] J. Sarek, M. Kvasnica, M. Urban, J. Klinot, and M. Hajduch, Correlation of cytotoxic activity of betulinines and their hydroxy analogues. *Bioorg. Med. Chem. Lett.* 15, 4196 (2005).

[26] E. L. Setti, and O. A. Mascaretti, Synthesis of (Pivaloyloxy)methyl 6α-Fluoro-penicillanate. *J. Chem. Soc., Perkin Trans. 1* 2059 (1988).

[27] ADHOC International Steering Committee. A randomized placebo-controlled trial of adefovir dipivoxyl in advanced HIV infection: the ADHOC trial. *HIV Medicine* 3, 229 (2002).

[28] Y. Zimra, A. Nudelman, R. Zhuk, E. Rabizadeh, M. Shaklai, A. Aviram, and A. Rephaeli, Uptake of pivaloyloxymethyl butyrate into leukemic cells and its intracellular esterase-catalyzed hydrolysis. *Journal of Cancer Research and Clinical Oncology* 126, 693 (2000).

[29] M. Urban, J. Sarek, M. Kvasnica, I. Tislerova, and M. Hajduch, Triterpenoid pyrazines and benzopyrazines with Cytotoxic Activity. *J. Nat. Prod.* 70, 526 (2007).

[30] J.-C. Leunis, and E. Couche, Betulonic and Betulinic Acid Derivatives. US 8586569, Nov. 19, 2013.

[31] D. Biedermann, B. Eignerova, M. Hajduch, and J. Sarek, Synthesis and Evaluation of Biological Activity of the Quarternary Ammonium Salts of Lupane-, Oleanane-, and Ursane- Type Acids. *Synthesis-Stuttgart* 22, 3839 (2010). (b) J. Sarek, D. Biedermann, and M. Hajduch, Triterpenoids Derivatives for treating cancer and pharmaceutical compositions containing them. CZ 301158, Sep. 12, 2009.

[32] S. Yang, N. Liang, H. Li, W. Xue, D. Hu, L. Jin, Q. Zhao, and S. Yang, Design, synthesis and biological evaluation of novel betulinic acid derivatives. *Chemistry Central Journal* 6, 141 (2012).

[33] H. Kommera, G. N. Kaluderovic, J. Kalbitz, B. Drager, and R. Paschke, Small structural changes of pentacyclic lupane type triterpenoid derivatives lead to significant differences in their anticancer properties. *Eur. J. Med. Chem.* 45, 3346 (2010).

[34] M. Willmann, V. Wacheck, J. Buckley, K. Nagy, J. Thalhammer, R. Paschke, T. Triche, B. Jansen, and E. Selzer, Characterization of NVX-207, a novel betulinic acid-derived anti-cancer compound. *European Journal of Clinical Investigation* 39, 384 (2009).

[35] C. Gauthier, J. Legault, S. Lavoie, S. Rondeau, S. Tremblay, and A. Pichette, Synthesis and Cytotoxicity of Bidesmosidic Betulin and Betulinic Acid Saponins. *J. Nat. Prod.* 72, 72 (2009).

[36] C. Genet, A. Strehle, C. Schmidt, G. Boudjelal, A. Lobstein, K. Schoonjans, M. Souchet, J. Auwerx, R. Saladin, and A. Wagner, Structure-Activity Relationship Study of Betulinic Acid, a Novel and Selective TGR5 Agonist, and Its Synthetic Derivatives: Potential Impact in Diabetes. *J. Med. Chem.* 53, 178 (2010).

[37] M. Evers, C. Poujade, F. Soler, Y. Ribeill, C. James, Y. Lelievre, J.-C. Gueguen, D. Reisdorf, I. Morize, R. Pauwels, E. De Clercq, Y. Henin, A. Bousseau, J.-F. Mayaux, J.-B. Le Pecq, and N. Dereu, Betulinic Acid Derivatives: A New Class of Human Immunodeficiency Virus Type 1 Specific Inhibitors with a New Mode of Action. *J. Med. Chem.* 39, 1056 (1996).

[38] K. Quian, S.-Y. Kim, H.-Y. Hung, L. Huang, C.-H. Chen, and K.-H. Lee, New betulinic acid derivatives as potent proteasome inhibitors. *Bioorg. Med. Chem. Lett.* 21, 5944 (2011).

[39] H.-J. Jeong, H.-B. Chai, S.-Y. Park, and D. S. H. L. Kim, Preparation of amino acid conjugates of betulinic acid with activity against human melanoma. *Bioorg. Med. Chem. Lett.* 9, 1201 (1999).

[40] V. Kumar, N. Rani, P. Aggarwal, V. K. Sanna, A. T. Singh, M. Jaggi, N. Joshi, P. K. Sharma, R. Irchhaiya, and A. C. Burman, Synthesis and cytotoxic activity of heterocyclic ring-substituted betulinic acid derivatives. *Bioorg. Med. Chem. Lett.* 18, 5058 (2008).

[41] N. Antimonova, N. I. Petrenko, E. E. Shults, E. E.; Y. F. Polienko, M. M. Shakirov, I. G. Irtegova, M. A. Pokrovskii, K. M. Sherman, I. A. Grigorev, A. G. Pokrovskii, and G. A. Tolstikov, Synthetic Transformations of Higher Triterpenoids XXX. Synthesis and Cytotoxic Activity of Betulonic Acid Amides with fragments of Nitroxyl Radicals. *Russ. J. Bioorg. Chem.* 39, 181 (2012).

[42] G. V. Giniyatullina, O. B. Kazakova, N. I. Medvedeva, I. V. Sorokina, N. A. Zhukova, T. G. Tolstikova, and G. A. Tolstikov, Synthesis of Aminopropylamino Derivatives of Betulinic and Oleanolic Acids. *Russ. J. Bioorg. Chem.* 39, 329 (2013).

[43] A. Tolmacheva, E. V. Igosheva, Y. B. Vikharev, V. V. Grishko, O. V. Savinova, E. I. Boreko, and V. F. Eremin, Synthesis and Biological Activity of Mono- and Diamides of 2,3-Secotriterpene Acids. *Russ. J. Bioorg. Chem.* 39, 186 (2013).

[44] U. V. Mallavadhani, A. Mahapatra, K. Jamil, and P. S. Reddy, Antibacterial Activity of Some Pentacyclic Triterpenes and Their Synthesized 3-*O*-lipophylic Chains. *Biological and Pharmacological Bulletin* 27, 1576 (2004).

[45] U. V. Mallavadhani, A. Mahapatra, S. S. Raja, and C. Manjula, Antifeedant Activity of Some Pentacyclic Triterpenes and Their Fatty Acid Ester Analogues. *J. Agric. Food Chem.* 51, 1952 (2003).

[46] M. de L. e Silva, J. P. David, L. C. R. C. Silva, R. A. F. Santos, J. M. David, L. S. Lima, P. S. Reis, and R. Fontana, Bioactive Oleanane, Lupane and Ursane Triterpene AcidDerivatives. *Molecules* 17, 12197 (2012).

[47] P. Reddy, A. B. Singh, A. Puri, A. K. Srivastava, and T. Narender, Synthesis of novel triterpenoid (lupeol) derivatives and their in vivo antihyperglycemic and antidyslipidemic activity. *Bioorg. Med. Chem. Lett.* 19, 4463 (2009).

[48] T. Tanachatchairatana, J. B. Bremner, R. Chokchaisiri, and A. Suksamrarn, Antimicrobial Activity of Cinnamate-Based Esters of the Triterpenes Betulinic, Oleanolic and Ursolic Acids. *Chem. Pharm. Bull.* 56, 194 (2008).

[49] O. B. Kozakova, N. I. Medvedeva, I. P. Baikova, G. A. Tolstikov, T. V. Lopatina, M. S. Yunusov, and L. Zaprutko, Synthesis of Triterpenoid Acylates: Effectivere Reproduction Inhibitors of Influenza A (H1N1) and Papilloma Viruses. *Russ. J. Bioorg. Chem.* 36, 771 (2010).

[50] M. Kvasnica, J. Sarek, E. Klinotova, P. Dzubak, and M. Hajduch, Synthesis of phthalates of betulinic acid and betulin with cytotoxic activity. *Bioorg. Med. Chem.* 13, 3447 (2005).

[51] O. B. Flekhter, L. T. Karachurina, V. V. Poroikov, L. P. Nigmatullina, L. A. Baltina, F. S. Zarudii, V. A. Davydova, L. V. Spirikhin, I. P. Baikova, F. Z. Galin, and G. A. Tolstikov, The Synthesis and Hepatoprotective Activity of Esters of the Lupane Group Triterpenoids. *Russ. J. Bioorg. Chem.* 26, 192 (2000).

[52] A. Baltina Jr., L. A. Baltina, R. M. Kondratenko, O. A. Plyasunova, S. A. Nepogodiev, and R. A. Field, Synthesis and Anti-HIV Activity of Triterpene 3-*O*-Galactopyranosides, Analogs of Glycyrrhizic Acid. *Chem. Nat. Compd.* 46, 576 (2010).

[53] J. Gao, X. Li, G. Gu, S. Liu, M. Cui, and H.-X. Lou, Facile synthesis of triterpenoid saponins bearing β-Glu/Gal-(1→3)-β-GluA methyl ester and their cytotoxic activities. *Bioorg. Med. Chem. Lett.* 22, 2396 (2012).

[54] J. Sarek, P. Spacilova, and M. Hajduch, Triterpenoid 2-deoxyglycosides, method of preparation thereof and use thereof as medicaments. US 8680251, Mar. 25, 2014.

[55] V. I. Grishkovets, Synthesis of Triterpenoid Sulfates Using the SO_3-Dimethyl Sulfoxide Complex. *Chem. Nat. Compd.* 35, 73 (1999).

[56] J. Castro, V. Richmond, C. Romero, M. S. Maier, A. Estevez-Braun, A. G. Ravelo, M. B. Faraoni, and A. P. Murray, Preparation, anticholinesterase activity and molecular docking of new lupane derivatives. *Bioorg. Med. Chem.* 22, 3341 (2014).

[57] C. Ma, N. Nakamura, and M. Hattori, Chemical Modification of Oleanane Type Triterpene and Their Inhibitory Activity against HIV-1 Protease Dimerization. *Chem. Pharm. Bull.* 48, 1681 (2000).

[58] H. Huang, P. Ho, K.-H. Lee, and C.-H. Chen, Synthesis and anti-HIV activity of bi-functional betulinic acid derivatives. *Bioorg. Med. Chem.* 14, 2279 (2006).

[59] D. Yu, Y. Sakurai, C.-H. Chen, F.-R. Chang, H. Li, Y. Kashiwada, and K.-H. Lee, Anti-Aids Agents 69. Moronic Acid and Other Triterpene Derivatives as Novel Anti-HIV Agents. *J. Med. Chem.* 49, 5462 (2006).

[60] Y. Liu, W.-X. Lu, M.-C. Yan, Y. Yu, T. Ikejima, and M.-S. Cheng, Synthesis and Tumor Cytotoxicity of Novel Amide Derivatives of □-Hederin. *Molecules* 15, 7871 (2010).

[61] A. Baltina Jr., O. V. Stolyarova, L. A. Baltina, R. M. Kondratenko, O. A. Plyasunova, and A. G. Pokrovskii, Synthesis and Antiviral Activity of 18α-Glycyrrhizic Acid and its Esters. *Pharmaceutical Chemistry Journal* 44, 299 (2010).

[62] J. M. Pezzuto, J. W. Kosmeder, Z.-Q. Xu, and N. E. Zhou, *Prodrugs of Betulinic Acid Derivatives for the Treatment of Cancer and HIV.* WO 02/16395 A1, Feb. 28, 2002.

In: Terpenoids and Squalene
Editor: Alanna R. Bates

ISBN: 978-1-63463-656-8

Chapter 3

TERPENOIDS OF THE SAHARAN MEDICINAL PLANTS *LAUNAEA* CASS. GENUS (ASTERACEAE) AND THEIR BIOLOGICAL ACTIVITIES

Abdelkrim Cheriti[1*] ***and Nasser Belboukhari*** [2]
[1]Phytochemistry and Organic Synthesis Laboratory, Algeria
[2]Bioactive Molecules and Chiral Separation Laboratory
University of Bechar, Algeria

ABSTRACT

Launaea Cass. is a small genus of the family Asteraceae (tribe Lactuceae, subtribe Sonchinae), comprises 54 species which 09 are presented in the flora of Algeria. Plants in the genus *Launaea* have been used ethnobotanically as bitter stomachic, for treating diarrhea, gastrointestinal tracts, as anti-inflammatory, for skin diseases, treatment of infected wounds, hepatic pains, children fever, as soporific, lactagogue, diuretic and used as insecticidal.

From a chemical point of view, only ten species of the genus *Launaea* Cass. have been subjected to previous phytochemical investigation, namely, *Launaea acanthoclada, L. arborescens, L. asplenifolia,L. capitata, L. cassiniana, L. mucronata, L. nudicaulis, L. pinnatifida, L. resedifolia* and *L. tenuiloba*. Different secondary

* Email: karimcheriti@yahoo.com.

metabolites including terpenoids, triterpenoid saponin, sesquiterpene lactones, steroids, polyphenolic compounds have been reported.

On the other hand, terpenoids are chemio-characteristic of Asteraceae family, including the *Launaea* genus, have been reported to have anti-inflammatory activities, anti-hyperlipidemia, hepatoprotection, antioxidant, cytoprotective, giving protection against cardiovascular disease, and certain forms of cancer.

We cover in this chapter the nature of terpenoids together with their biological activities to provide a comprehensive compilation of these compounds from the genus *Launaea Cass.*, especially those growing in Algerian Sahara and used as medicinal plants, namely: *Launaea arborescens*, *L. nudicaulis and L. residifolia*.

Keywords: Asteraceae, *Launaea* Cass., Terpenoids, biological activities, Sahara

1. Introduction

Medicinal plants in traditional pharmacopeia have been employed for the treatment and management of various ailments since the beginning of human civilization and continue to provide mankind with new remedies, such as, the oldest known medicinal systems of the world: Ayurveda, Arabian medicine, Chinese and Kempo medicine [1]. The Materia Medica of Ibn Baytar (Andalusia, Spain, 1197–1248), is one of the oldest documents that describe the use of natural products for healing diseases in the mediterranean area. It listing about 1400 drugs, medicinal plants and foods, and their uses with 350 new plant species as medicinal herbs for treating human diseases [2-4]. Although, medicinal plant therapy is based on the empirical findings of hundreds and thousands of years. Thus, The World Health Organization (WHO) has recognized the potential utility of traditional remedies and strives to preserve the primary health care involving medicinal plants [1, 5]. One of the most efficient ways of finding new bioactive compounds is collecting data on the use of medicinal plants in traditional pharmacopeia. Nearly 50,000 species of higher plants have been used for medicinal purposes. In systems of traditional healing, major pharmaceutical drugs have been either derived from or patterned after compounds from biological diversity [6] and are very important to conduct research on and they can be a source of new compounds [7]. In phytotherapy the use of new plant origin bioactive compounds with the potential for chemical modification, which will broaden phytomedical importance, the pharmacological profiles of these bioactive compounds are screened using new research equipment and new technology [8, 9]. Natural products and their derivatives represent more than 50% of all the drugs in clinical use in the world and in which higher plants contribute to no less than 25% [5].

Recentely Newman and Gragg [10] analyzed the origin of 1073 new chemical entities and conlued that 2/3 of the newly introduced drugs originated in some manner from a natural product: 6% of these drugs were natural products themselves, 28% were derivatives of natural products, 14% were characterized as mimics of natural products, and 16% as synthetics whose pharmacophore was derived from a natural product. Only a small part of the 400,000 vegetable species known have so far been investigated for their phytochemical and pharmacological aspects, and each species could contain up to several thousands of different components [11]. Among these bioactive compounds, terpenoids have been utilized since ancient times and old cultures by using essential oils for many purposes. Actually, these natural products have become of widespread importance in pharmacology, perfumery, beverages, foods and detergents. Terpenoids comprise the largest class of natural products displaying a wide range of biological and pharmacological activities. They can act as defensive substances in plants (allomones), can be used by plants to deter herbivores or to inform conspecifics, or attract natural enemies of herbivores (synomones). Plant hormones are often derivatives of terpenoids, such as cytokinins, gibberellins and abscisic acid [12].

Terpenoids are structurally diverse natural products with so many functions that originate from squalene or oxidosqualene by a series of intramolecular condensation reactions and can have a simple aliphatic structure, or a cyclic one. Monocyclic, bicyclic, tricyclic and polycyclic structures [6]. So, the diversity of structures obtained and the different therapeutic activities shown by the natural products make the isolation, identification, synthesis and biosynthesis of new natural terpenoids a field of enormous interest.

It is well known that species from Asteraceae family are used as natural remedies such as: *Anthemis arvensis* L. (anti-inflammatory, emetic, sedative), *Artemisia arborescens* L. (digestive, stimulant, expectorant), *Calendula arvensis* (antispasmodic, burns, diuretic, disinfectant and vulnerary), *Cichorium intybus* L. (blood purification, arteriosclerosis, anti-arthritis, anti-spasmodic, digestive, hypotensive, aperitif and laxative) and *Helychrysum microphyllum* Willd. (expectorant) [1].

Launaea Cass., named after naturalist and lawyer Jean Claude Mien Mordant de Launay (1750-1816), is a small genus of the family Asteraceae (tribe Lactuceae, subtribe Sonchinae) mainly distributed in the South Mediterranean, Africa and SW Asia. They are perennial to annual herbs, small shrubs or sub shrubs. Many of its plants are still used as medicinal plants, particularly in folk medicine and used as bitter stomachic, for treating diarrhea, gastrointestinal tracts, as anti-inflammatory, for skin diseases, treatment of infected wounds, hepatic pains, children fever, as soporific, lactagogue, diuretic and as insecticidal. Additionally, crude extracts of some species have been reported to exhibit antibacterial, antiparasitic, antioxidant, cytotoxic, neuropharmacological and insecticidal activities. From a chemical point of view, only ten species of the genus *Launaea* Cass. have been subjected to previous phytochemical investigation, namely, *Launaea acanthoclada, L. arborescens, L. asplenifolia, L. capitata, L. cassiniana, L. mucronata, L. nudicaulis, L. pinnatifida, L. resedifolia* and *L. tenuiloba*. Different secondary metabolites including terpenoids, steroids, triterpenoid saponin, sesquiterpene lactones, coumarins, flavonoids, flavone glycosides and phenolic compounds have been reported [1].

We present in this chapter the nature of terpenoids together with their biological activities to provide a comprehensive compilation of these compounds from the genus *Launaea Cass.* (Tribe Lactuceae), especially those growing in Algerian Sahara and used as medicinal plants.

2. Ethnobotanical and Bioactivity of the Genus *Launaea* Cass. (Tribe Lactuceae)

The tribe Lactuceae (Cichorieae, Asteraceae family) comprises 98 genera and more than 1550 species. The milky latex and the floral structure make the tribe easily distinguishable from all other Asteraceae [13]. The genus *Launaea* Cass. belongs to the tribe Lactuceae of the Asteraceae family and contains about 54 species, most of which are adapted to dry, saline and sandy habits [14]. Plants of this genus have several rows of stems, hairless leaves incised into lobes that are themselves lined with white teeth, membranous scales on the edges, yellow ligules, and elongated chain, prismatic or slightly flattened [1].

According to classification system on flowering plants [15], the classification hierarchy of the genus *Launaea* can be tracked as follows:

Kingdom:	Plantae
Division:	Angiosperms
Class:	Eudicots
Subclass:	Asterids
Order:	Asterales
Family:	Asteraceae
Subfamily:	Cichorioideae
Tribe:	Lactuceae (Cichorieae)
Sub-tribe:	Sonchinae
Genus:	*Launaea*

The genus *Launaea*, is represented in the flora of Algeria by nine species including five endemics of North Africa: *L. angustifolia*, *L. quercifolia*, and *L. cassiniana* are the endemic plants of the North Africa, with limited distribution [14, 16, 17], whereas *L. acanthoclada* and *L. arborescens* are two endemic plants of the north-west of Africa. The other four species *L. nudicaulis* and *L. residifolia* sprout in Algeria and Tunisia Mediterranean Sea, whereas *L. glomerata* and *L. mucronata* grow in the Saharan Atlas [17].

Algeria with its large area and diversified climate has a varied flora, which is a source of rich and abundant medical matter and, in particular, Sahara part constitutes an important reservoir of many plants which have not been investigated until today. Among this flora, some *Launaea* plants have been used in the traditional medicine [18-20]. Species of the genus *Launaea* are widely applied in traditional folk medicine throughout their areas of distribution. Many of them are used in folk medicine as bitter stomachic, anti-tumour, insecticides and against skin diseases. Three of this species are used in Algerian Sahara ethnopharmacopea as medicinal plants, *Launaea Nudicaulis, Launaea residifolia and Launaea arborescens*which is endemic to southwest Algeria and south east Morocco [18].

Launaea arborescens (Batt.) Murb, (syn. *Zollikoferia spinosa* DC, *Zollikoferia arborescens* Batt.) is an almost leafless, xerophilous, perennial spiny shrub, 40-120 cm. high, with typical zig-zag shaped stems (figure 1). The young stems are green, glabrous and erect. The older ones become tough spines. The leaves are narrow and dissected in small lobes, evergreen at the base but shed after flowering from the stems. The flowers are yellow, and abundant flowering occurs from March to June, but flowers and achenes are produced throughout the year. The roots are very deep, the leaves and stems have white latex which is similar in appearance to milk (thus the vernacular name "Oum loubina") [14, 17-19].

Figure 1. The endemic *Launaea arborescens* (Batt.) Algerian Sahara.

Launaea arborescens (Batt) commonly used in popular medicine as an antidiarrhoic and antispasmodic, to relieve fever, and as a vermifuge in children. The latex is applied locally to cure sore throats and in the treatment of furuncles. The powdered root mixed with *Artemisia herba-alba* is taken for diabetes. The plant is appreciated by livestock, mainly by camel [4, 18-21].

Launaea nudicaulis (syn. *Zollikoferia nudicaulis* Boiss., *Sonchus divaricatus* Desf., Vernacular name: Reghama) is used in the traditional medicine to treat gastric burns, pain of stomach, constipation, to stop excessive bleeding after childbirth, inthetreatmentofitchesof skin and eczema and the leaves are put on the heads of children to cure fever [18].

Launaea residifolia (L.) Kuntze (syn. *Zollikoferia resedifolia* Coss., *Scorzonera resedifolia*, Vernacular name: Lemkar) is a perennial stoloniferous herb widely distributed in the arid regions of the Mediterranean. This medicinal plant is used in folkloric medicine mainly for the treatment of hepatic pains [18].

Antibacterial, antifungal and allelopathic potential activities have been proven for many species of *Launaea*. In an antibacterial assay against *Bacillus subtilis* the extracts of *Launaea nudicaulis* and *Launaea residifolia* showed

18.5 and 20.5 mm zones of inhibition, respectively, as determined by the disc diffusion method. The antifungal activity against *Aspergillus* spp. was determined by measuring the linear growth in slants on 4th day of incubation. Methanol extracts of *L. nudicaulis* and *L. residifolia* were active at 0.209 mg/ml levels exhibiting 45±6 mm and 37±6 mm linear growth which decreased to 22±5 mm and 28±4 mm, respectively, at 0.838 mg/ml concentration [22].

As a part of our works on medicinal plants of Algerian Sahara, we have reported the antibacterial activity of extracts from *Launaea Arborescens* and *L. nudicaulis* which are widely distributed in the south west of Algeria. The methanol extract of the aerial part of *L. nudicaulis* showed high activities against *Candida albicans, Escherichia coli*, *Staphylococcus aureus* and *Pseudomonas aeruginosa*. The highest inhibition observed in *S. aureus*, a human pathogen, explains the use of this plant against a number of infections for generations. Very interesting antifungal activity against *Candida albicans* and *Saccharomyces cerevisae* and antibacterial activity against *Staphylococcus aureus, Escherichia coli*, *Pseudomonas aeruginosa* and *Klebsiella entrecocus* have been reported for the methanol extract of *Launaea Arborescens* [19, 23].

Hydroalcoholic extract from aerial parts of *Launaea arborescens* was evaluated for acute and subacute toxicity in *Swiss* mice after ingestions of the extract. The LD50 of the extract is higher than 2.75 g/kg and the subacutetreatment did not shows any change in corporal weight and haematological parameters, which suggest that the plant seems to be destituted of toxic effects in mice [24]. Aerial part and roots of *Launaea arborescens* were used to evaluate their extracts for antifungal activity against *Fusarium oxysporum* f. sp. albedinis Foa. The antifungal test was conducted using disc diffusion technique and relative virulence (RV) test on potato tuber tissue. For both tests, four extract quantities were used (200, 400, 800 and 1,600µg). The relative virulence was presented as necrotic tissue weight (mg) of potato tuber tissue. Among all solvents, methanol had the best extraction yield (mean: 6.35%, minimum: 2.27%, maximum: 9.80%) [25].

The in vitro antioxidant activity of *Launaea nudicaulis* extracts was investigated with DPPH radical scavenging assay. The quantitative evaluation of DPPH scavenging activity showed that n-BuOH and EtOAc extracts are the most active extracts with a percentage of antiradical activity of 89,62% and 71,57% respectively [26].

The ethanol extract of *L. resedifolia* showed neuropharmacological properties in animal models. The extract exhibited an inhibitory effect on the locomotor activity of mice in the open field test, an anti-nociceptive effect by increasing the hot plate reaction time in the hot plate test, and an anti-inflammatory activity in the carrageen-induced paw oedema. This finding has demonstrated that the extract of *L. resedifolia* possesses sedative, analgesic and anti-inflammatory properties, and some effect on body weight. The anti-inflammatory effect of the plant was found to be as active as the prototype non-steroidal anti-inflammatory drug (NSAID) aspirin [27].

3. TERPENOIDS OF THE GENUS *LAUNAEA* CASS. (TRIBE LACTUCEAE)

The biodiversity of metabolite products isolated from Asteraceae makes this family an important phytochemical and commercial source. Several phytochemical studies of some genera of *Lactuceae* tribe (Cichorieae) revealed to be rich in secondary metabolites, specifically sesquiterpene lactones exhibiting the eudesmane, germacrane and guaiane carbon framework. A total of 360 sesquiterpene lactones and related compounds have been isolated from 139 taxa belonging to 31 different genera of the *Lactuceae*. Studies realized for these genera revealed that most sesquiterpenoids within the Cichorieae belong to the guaianolide class, particularly: 92 representatives of costus lactone type, 75 compounds of lactucin type, and 29 representatives of hieracin type [28, 29].

On the other hand, triterpenoids and flavonoids chemio-characteristic of Asteraceae family, including the *Launaea* genus, have been reported to have anti-inflammatory activities, anti-hyperlipidemia, hepatoprotection, antioxidant, cytoprotective, giving protection against cardiovascular disease, and certain forms of cancer [30, 31].

3.1. Terpenoids Isolated from the 2nd Group of Lactuceae Tribe

Based on the similarity of their sesquiterpenes profiles, Zidorn [28] grouped the 31 genera of the *Lactuceae* into seven main clusters and classified Launaea with the 2ed group characterized by the prevalence of guaianolides, formed by 11 genera, sub-divided into four sub-groups:
a) *Scorzoneroides*; b) *Notoseris, Lactuca* and *Cichorium*; c) *Launaea, Crepidiastrum*, *Reichardia* and *Cicerbita;* d) *Taraxacum*, *Helminthotheca* and *Hypochaeris*.

Actually more than 20,000 triterpenoids have been identified from the various parts of medicinal plants. These natural compounds are constituents of the cell membrane, and they play important roles in cell membrane defense. The diversity of skeletal structures, isolation technic, biological and pharmacological activities are the driving forces for research work on these natural compounds [32]

3.1.1. Triterpenoids

The majority of the triterpenoides are pentacyclic and belong to lupane, oleanane, gammacerane and ursane groups, with some tetracyclic compounds such as β-amyrin **1**, β-amyrin acetate **2**, taraxerol **3**, taraxeryl acetate **4**, taraxasterol **5**, taraxasterone **6**, taraxasteryl acetate **7**, ψ-taraxasteryl derivatives **8**, **9**, α-amyrin derivatives **10**, **11**, lupeol **12**, lupenone **13**, and lupenyl acetate **14** [33, 34].

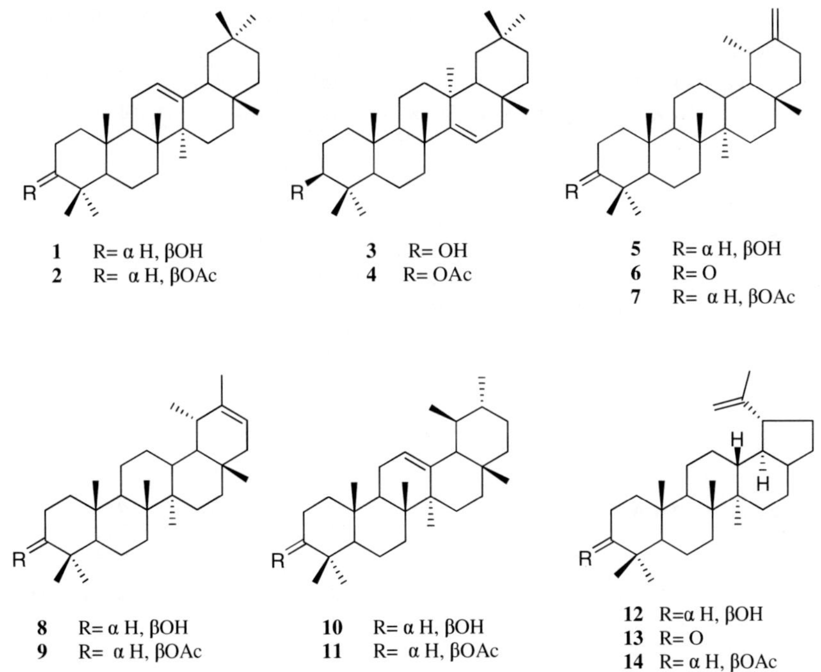

3.1.2. Sesquiterpenoids

More than 4000 sesquiterpenoids structures with around 30 different skeletal types have so far been reported from several tribes of Asteraceae family including the Cichorieae tribe. These natural compounds are responsible for allergic contact dermatitis and exhibit a wide range of bioactivities which include plant growth regulation and antimicrobial activity. Also they are used as schistosomicidal and insect feeding-deterrent agents. In addition, they provoke the toxicity for certain cancer cell lines by inhibition of nuclear DNA synthesis, especially the enzymatic activity in tumour cells of DNA polymerase and thymidylate synthetase [35-37]. Due to their chemo-diversity, the sesquiterpene lactones are the most suitable class of natural products for chemo-systematic studies within the family [35, 38].

Costus lactone type guaianolides such as dehydrocostruslactone **15**, ixerisoside B **16**, C **17** and D **18**, scorzoside **19**, zaluzanin C **20**,glucozaluzanin C **21,** 11 β,13-dihydrozaluzanin C **22**,8β-hydroxy-4 β,15-dihydrozaluzanin C **23** and prenantheside C **24** [33, 39]. Lactucin type guaianolides, Lactucin **25**, 8-O-acetate Lactucin **26**, Crepidiaside A **27**, 11β, 13dihydrolactucin **28**, 8-O-acetate,11β,13dihydrolactucin**29** and8-Deoxylactucin **30.** The eudesmane derivatives santamarin **31**, ixerisoside E **32**, lactuside D **33**, sonchuside C **34** and artesin **35**[33, 40], costinolide type germacranolides such as picriside B **36**, C **37**, sonchuside A **38**, B **39** and cichoerioside C **40**,[41-43], and melampolides type, lactulide A **41**, lactuside A **42** and B **43**[33, 39, 44, 45].

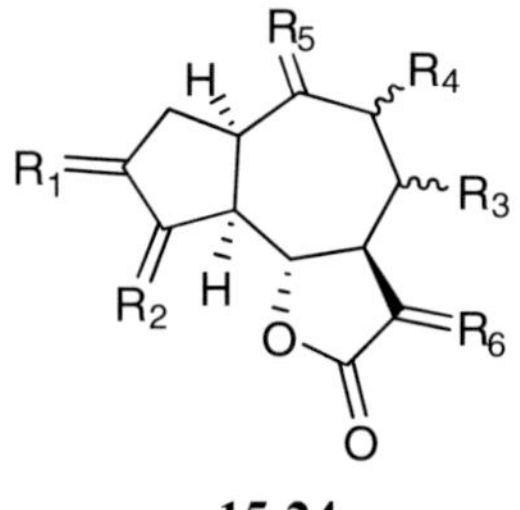

15-24

	R_1	R_2	R_3	R_4	R_5	R_6
15	H,H	CH_2	H	H	CH_2	CH_2
16	α H,β OGlc	α CH_3, βH	β OH	H	CH_2	CH_2
17	α H,β OGlc	CH_2	H	H	CH_2	αCH_3, βH
18	H,H	CH_2	H	α OGlc	CH_2	CH_2
19	H,H	CH_2	H	α OGlc	CH_2	α CH_3, β
20	α H,β OH	CH_2	H	H	CH_2	CH_2
21	α H,β OGlc	CH_2	H	H	CH_2	CH_2
22	α H,β OH	CH_2	H	H	CH_2	α CH_3, βH
23	α H,β OH	α CH_3, βH	β OH	H	CH_2	CH_2
24	α H,β OGlc	CH_2	α OH	H	CH_2	α CH_3, Bh

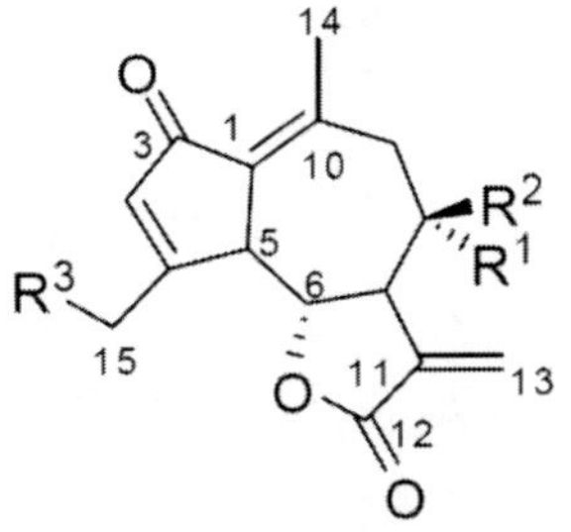

25-27

25 $R_1 = R_3$=OH , R_2=H
26 R_1 = OAc , R_2=H , R_3=OH
27 $R_1 = R_2$=H , R_3=O-glc

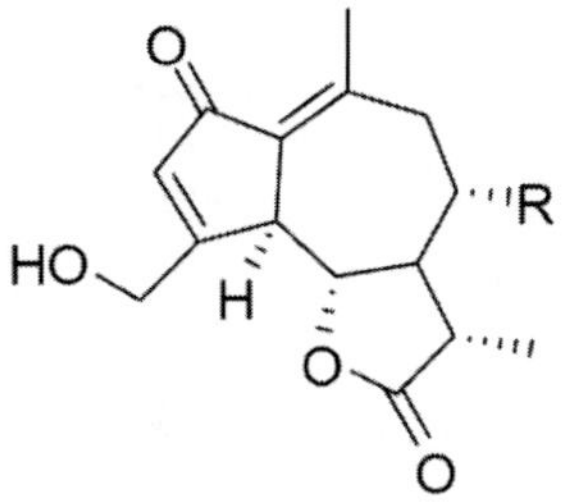

28-30

28 R=OH
29 R = OAc
30 R=H

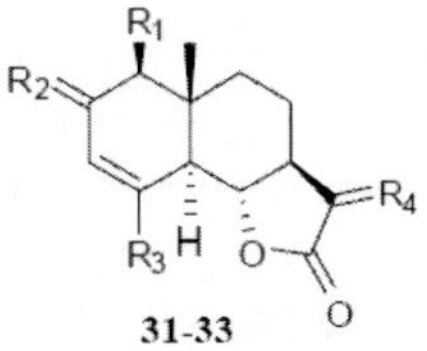

31-33

	R_1	R_2	R_3	R_4
31	OH	H ,H	CH_3	CH_2
32	OH	H ,H	CH_2OGlc	CH_2
33	O-PPA	H ,H	CH_2OGlc	CH_2

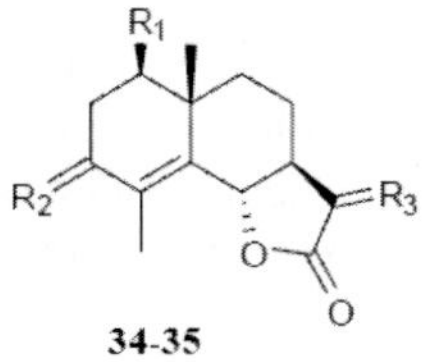

34-35

	R_1	R_2	R_3
34	H	α H, β OGlc	α CH_3, β H
35	OH	H, H	α CH_3, β H

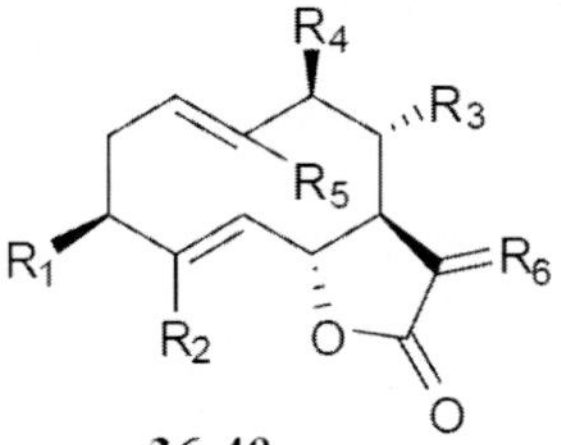

36-40

	R_1	R_2	R_3	R_4	R_5	R_6
36	H	CH_2OGlc	H	H	CH_3	CH_2
37	OGlc	CH_3	H	H	CH_3	CH_2
38	O-Glc	CH_3	H	H	CH_3	α CH_3, β H
39	OGlc	CH_3	H	O-PMP	CH_3	CH_2
40	OGlc	CH_3	OH	H	CH_3	α CH_3, β H

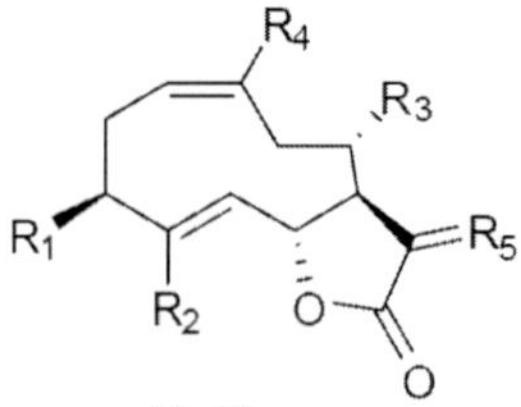

41-42

	R_1	R_2	R_3	R_4	R_5
41	OH	CH_3	H	CHO	α CH_3 ,βH
42	OGlc	CH_3	H	CHO	α CH_3 ,βH
43	OGlc	CH_3	H	CH_2OH	α CH_3 ,βH

3.2. Terpenoids Isolated from the Saharan *Launaea* Genus

Different secondary metabolites have been identified from the genus *Launaea*. In addition, few sesquiterpene lactones have been reported from various species of this genus and the occurrence of flavones glycosides is remarkable. The first works in phytochemistry on species of the genus *Launaea* was started in 1969 by Prabhu and Venkateswarlu [46], when they isolated from leaves and roots of *Launaea pinnatifida* two compounds Taraxasterol **5** and Taraxasterly acetate **7**. Twenty year ago, in 1989, Gupta et al. [47] investigated *Launaea asplenifolia* and isolated, taraxasterol, taraxasterone, taraxasteryl acetate. Triterpenoids taraxasterol and taraxerol exhibited potent anti-tumourpromoting activity in carcinogenesis tests of mouse skin (induced by a chemical initiator and a promoter). In addition they had an inhibitory effect on mouse spontaneous mammary tumours, indicating that terpenoids with anti-tumour effects can be potentially anti-carcinogenic [48].

3.2.1. Launaea arborescens

Chemical data on this species are scarce in literature and few published papers describe phenolic components of the plant. In their studies on *Launaea* genus from Spain including *L. arborescens*, Giner *et al.* [49] isolated common phenolic compounds (luteolin, luteolin-7-*O*- glucoside, luteolin-7-*O*-rhamnoside, esculetinand cichoriin)

In our first studie on the phytochemical study of the Algerian sample of *Launaea arborescens* collected from the Sahara [50], we have described the isolation from the methanol extract of the aerial parts of this specie, four compounds, two flavonoids, one diepoxylignanand the diterpene, methyl-15,16-epoxy-12-oxo-8(17),13(16),14-ent-labdatrien-19-oate **44.**

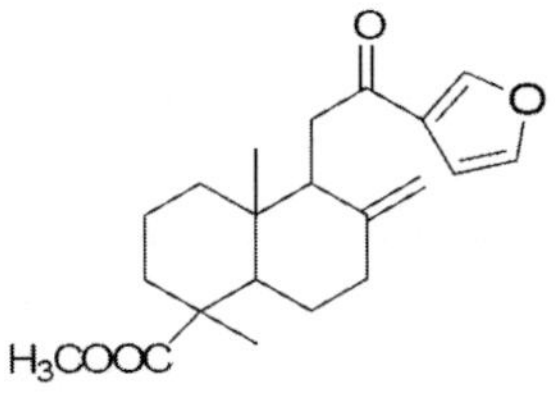

H_3COOC

44

A new 3β-hydroxy-11α-ethoxy-olean-12-ene **45** was isolated from the aerial parts extract of *Launaea arborescens*, to ghether with a series of oleanane and ursane triterpenes such as: 3β-hydroxy-11α-methoxyolean-12-ene **46**, 3β-11α-dihydroxy-olean-12-ene **47**, 3-β-hydroxy-11-oxo-olean-12-ene **48**, oleana-9(11):12-dien-3β-ol **49**, 3β-hydroxy-11α-ethoxyurs-12-ene **50**, 3β-hydroxy-11α-methoxyurs-12-ene **51**, 3β-11α-dihydroxy-urs-12-ene **52** and 3-β-hydroxy-11-oxo-urs-12-ene **53** [51].

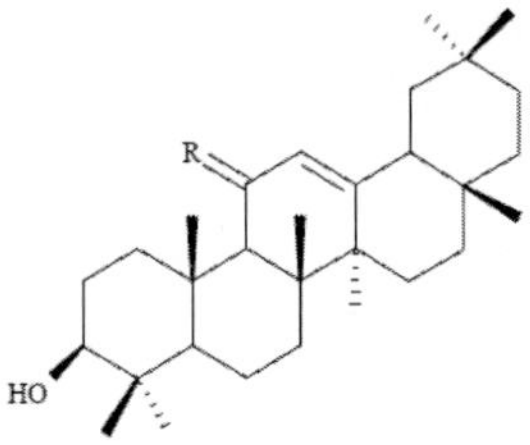

R

HO

45- 47

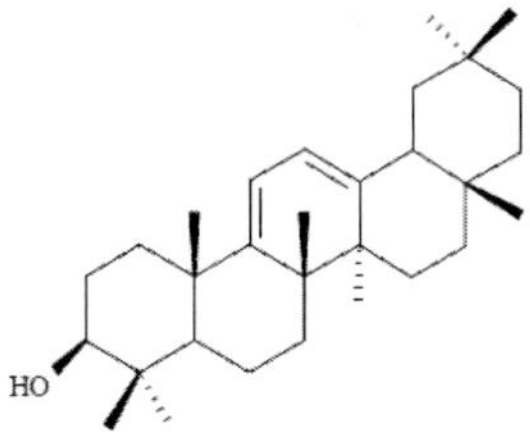

HO

49

45 R = α OCH_2CH_3 , βH
46 R = α OCH_3, βH
47 R = α OH, βH
48 R = O

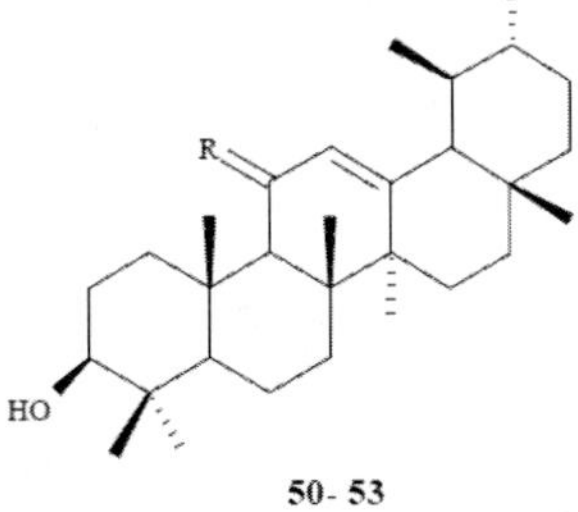

R

HO

50- 53

50 R = α OCH_2CH_3 , βH
51 R = α OCH_3, βH
52 R = α OH, βH
53 R = O

Fom the ethyl acetate extract of the roots of this specie was characterized a series of sesquiterpene derivatives exhibiting eudesmane, guaiane and germacrane skeletons, and identified as 9α-hydroxy-11β,13-dihydro-3-epi-zaluzanin C **54**, 9α-hydroxy-4α,15-dihydrozaluzanin C **55**, 3β,14-dihydroxy costunolide-3-O-β-glycopyranoside **56**, 3β,14-dihydroxycostunolide -3-O-β-glucopyranosyl-14-O-p-hydroxyphenylacetate **57** and a sesquiterpenoid sulphate, 8-deoxy-15-(3'-hydroxy-2'-methylpropanoyl)-lactucin-3'sulfate **58** [51].

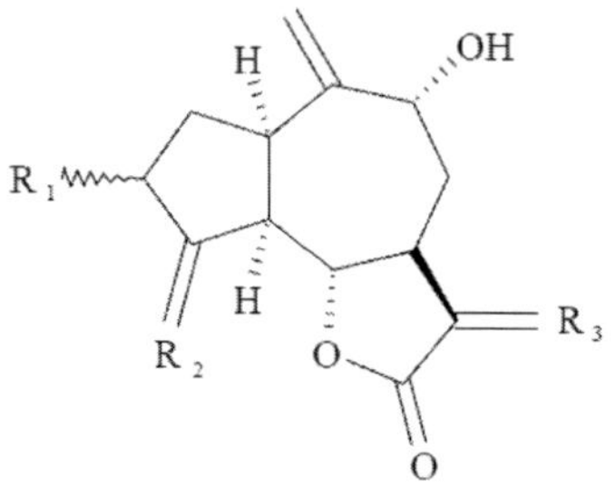

54-55

	R_1	R_2	R_3
54	α OH	CH_2	α CH_3 ,βH
55	β OH	β CH_3	CH_2

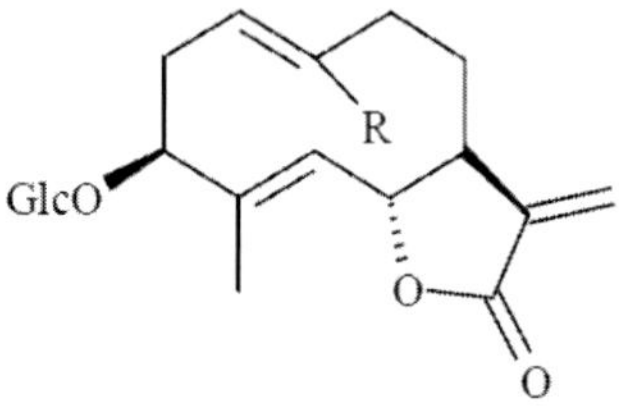

56-57

56 R= CH_2OH

57 R=

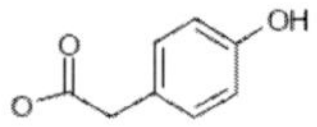

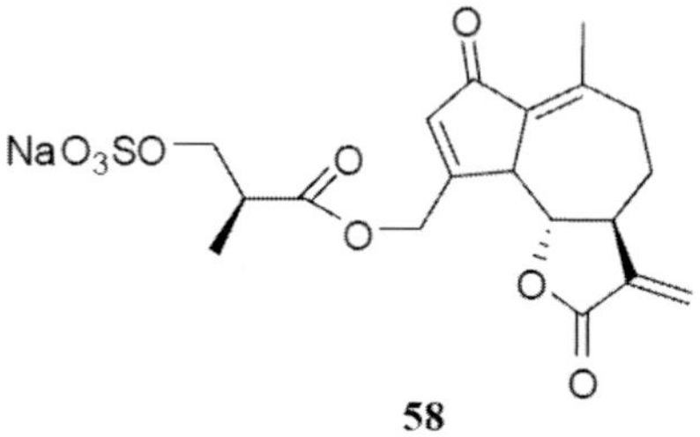

58

All identified terpenoids **45** and **54-57** were tested for both antifungal and antibacterial activity at a concentration of 5 μg/ml. No growth inhibition was exhibited on *Candida albicans* as well as on *Escherichia coli* and *Staphylococcus aureus* [51].

The hydrodistillation of the aeriel part of *Launaea arborescens* gave a green yellowish oil in an yield of 0.07% from dried material. Seventeen compounds were identified, representing 84.96% of the total oil. The essential oil of *L. arborescens* was a mixture of different substances, including oxygen-containing monoterpenes, alcohols, aldehydes, and esters. Esters were the dominant group in the oil (58.24%), the terpenoid portion consisted of two oxygenated monoterpenes accounting for 7.24% of the oil. We also found aldehydes in considerable amounts (16.09%) [52].

3.2.2. Launaea Nudicaulis

The light petroleum extract of *Launaea nudicaulis* leads to the characterization of some $\Delta^7$and Δ^5phytosterols: β-sitosterol, brassicasterol, campesterol, stigmasterol, fucosterol, 24β-$\boldsymbol{\Delta}^7$-ergosten-3β-ol and stigmasta-7,24(28)-dien-3–ol [53]. Detailed chemical investigation of *Launaea nudicaulis* yielded some triterpenes such as taraxasterol **5**, ψ- taraxasterol **8**,3β- taraxerol **3**, α- amyrin **10**, and lupeol **12** [54].

Two new ursene type triterpenes, nudicauline A **59**, and nudicauline B **60** have been isolated from the aerial parts of this specie, along with olean-11,13(18)-diene**61**,3β-hydroxy-13(28)-epoxy-urs-11-ene**62**and3-keto-13(28)-epoxy-urs-11-ene **63** [55].

RO

59-60

59 R= OH
60 R = Ac

HO

61

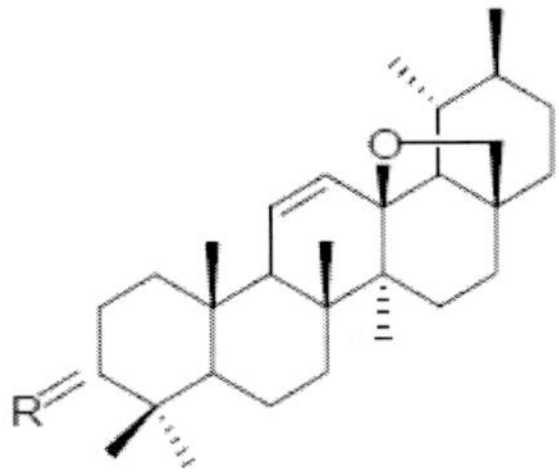

R

62-63

62 R= α H, β OH
63 R = O

Recentely, ethyl acetate soluble fraction of methanolic extract of *Launaea nudicaulis* was subjected to chromatographic purification to get new quinic acid and flavone glycoside derivative with the sesquiterpene lactone nudicholoid **64**, which exhibited a moderate inhibitory activity against the enzyme butyrylcholinesterase with an IC50 value of 88.3 mM [56].

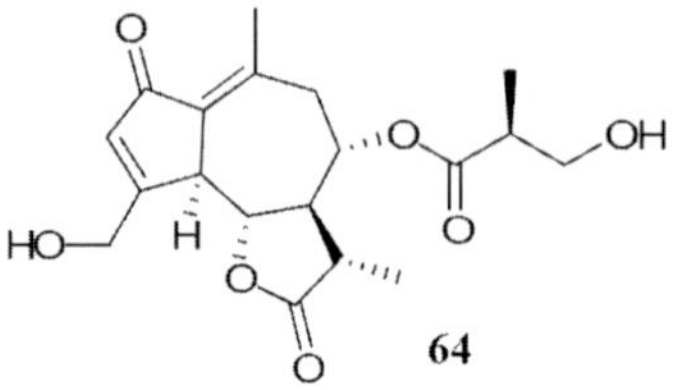

64

3.2.3. Launaea residifolia (L.)

Chemical studie of the plant led to the isolation of triterpenes α-amyrin **10**, lupeol **12**, lupeol acetate **14** and their epimer moretenol together with the Δ^7-stigmasterol [57].

On the other hand, the chemical composition of essential oils from this specie (0.9%) has been identified using the ordinary GC-MS technique. Nineteen compounds of essential oil of *L. residifolia* L. were identified representing 86.68% of the total oil. The compounds were identified by spectral comparison to be mainly esters, alcohols, ketones, and terpenes [58].

CONCLUSION

The genus *Launaea* (Lactuceae tribe, Asteraceae family) has great importance due to its ethnobotanics, phytochemistry and biological activity, and it is a promising source of various secondary metabolites including sequiterpenoids, triterpenoids. Some of these isolates compounds have been found to exhibit various biological activities.

This chapter covered presents information on the importance of the terpenoids as well as their biological significance of the members of this genus, especially the species growing in Algerian Sahara namely: *Launaea arborescens*, *Launaea nudicaulis and Launaea residifolia* (L.). The given information can be the base for undertaking future research.

REFERENCES

[1] Cheriti A., Belboukhari M., Belboukhari N., Djeradi H., *Current Topics in Phytochem.* (2012), 11, 67.

[2] Ibn al-Baytar Dhiya al-Din, 12th century, *Kitāb al-jāmiʽ li-mufradāt al-adwiya wa al-aghdhiya* (Compendium on Simple Medicaments and Foods), (1992), Ed. Dar Kotob Elmia, Libanon

[3] Leclerc L., *Traité des simples de ʽAbd Allāh ibn Aḥmad Ibn al-Bayṭār*, Traduction, (1877) Impr. Nationale, Paris.

[4] Bellakhdar J., *La pharmacopée marocaine traditionnelle. Médecine arabe ancienne et savoirs populaires*, (1997), IBIS Press.

[5] Gurib-Fakim A., *Traditions of yesterday and drugs of tomorrow. Molecular Aspects of Medicine.* Special Edition. (2006), Elsevier Publications, UK.

[6] Bisset N., *Herbal drugs and phytopharmaceuticals. A handbook for practice on a scientific basis.*, (1994), Ed. Medpharm. Sc. Publishers, Sttutgarts-CRC Press, Boca Raton.

[7] Newman, D. J., Cragg, G. M.,*Journal of Natural Products,* (2007), 70, 461.

[8] Cordell, G., Colvard, M.., *J. Ethnopharmacol.*, (2005), 100, 5.

[9] Yaniv Z. and Bachrach, U., *Handbook of Medicinal Plants*, (2005), Harworth Press, Inc. New York, London.

[10] Newman D.J., Cragg, G.M., *Journal of Natural Products*, (2012), 75, 311.

[11] Hostettmann, K., Potterat, O., Wolfender, J.-L., *Chimia*, (1998),52, 10.

[12] Harrewijn P., Van Oosten A. M., Piron P. G.M., *Natural terpenoids as messengers: a multidisciplinary study oftheir production, biological functions, and practical applications*, (2000), Kluwer Academic Publishers.

[13] Bremer. K., *Asteraceae: Cladistics and Classification*, (1994), Timber Press, Portland.

[14] Ozenda, P., *Flore et Végétation du Sahara*, (2004), CNRS, Paris.

[15] Funk, V.A., Susanna, A., Stuessy, T.F., Bayer, R.J., *Systematics, Evolution and Biogeography of Compositae,* (2009), International Association for Plant Taxonomy, Vienna, Austria.

[16] Cheriti A., *Plantes Médicinales de la Région de Bechar, Sud Ouest Algérie: Etude Ethnopharmacologique*, (2000), Rapport CRSTRA‘, Algeria.

[17] Quezel, P., Santa, S., *Nouvelle Flore d'Algérie et des Régions Désertiques Méridionales*, (1963), vol. 1–2. CNRS, Paris.

[18] a) Cheriti, A., Belboukhari, N., Sekkoum, K.. & Hacini, S., *J. Algerien reg. Arides*,(2006),5,7. b) Sekkoum K., Belboukhari N., Cheriti A., *Asian Pacific Journal of Tropical Biomedicine*, (2014), 4(4), 267.

[19] a) Cheriti, A., Belboukhari, N., Hacini, S., *Annales Univ. de Bechar*, (2001), 1. 4. b) Belboukhari N., Cheriti A. andRoussel C., *eCAM*,(2007), 4(S1), 55. c) Bourmita Y., Cheriti A., Ould El Hadj M. D., Mahmoudi K., (2013), *Journal of Entomology* 10, 1.

[20] a) Belboukhari,N., Cheriti A., *Elec. J. Environ., Agron., Food Chem.*, (2008),7(14), 2749. b) Cheriti A., Belboukhari N., Hacini S., *PhytoChem & BioSub Journal*, (2013), 7(2), 52.

[21] Chehma, A., *Catalogue des plantes spontanées du Sahara septentrional Algérien*, (2006), Ed. Dar ElHouda, Ain Mila, Algeria.

[22] Rashid, S., Ashraf M., Bibi S., Anjum R., *Pak. J. Bio. Sci.,* (2000), 3, 630.

[23] Belboukhari, N, Cheriti A., *Pak. J. Biolog. Sci.,* (2006), 9(1),1.

[24] Salah Ramadan B., Cheriti A., Belboukhari N., Zaouani M., *Annales Univ Bechar*, (2008), 4, 48.

[25] Boulenouar N., Marouf A., Cheriti A., Belboukhari N., *J. Agr. Sci. Tech.,*(2012), 14, 659.

[26] Belboukhari M., Cheriti A., Belboukhari N., *Ann. Fac. Sci.,* (2014),6(1), 52.

[27] Auzi, A.A., Hawisa, N.T., Sherif, F.M., Sarker, S.D., *Rev. Bras. farmacogn.*, (2007), 17, 160.

[28] a) Zidorn, C., *Biochem. System. Ecol.*, (2006), 34, 144. b) Zidorn, C., *Phytochemistry* (2008),69, 2270.

[29] Zidorn, C., Ellmerer, E.-P., Heller, W., Jöhrer, K., Frommberger, M., Greil, R., Guggenberger, M., Ongania, K.-H., Stuppner, H., *Z. Naturforsch.*, (2007),62b, 132

[30] Lie, J., *J. Ethnopharmacol.*, (1995),49, 57.

[31] Cheriti, A., Babadjamian, A., Balansard, G.,*Journal of Natural Products*, (1994),57, 1160.

[32] a) ConnollyJ.D., Hill R.A., *Nat. Prod. Rep.* (2010),27, 79. b)Liby K.T., Yore M.M., Sporn M.B., *Nat. Rev. Cancer*, (2007),7, 357.

[33] Kisiel, W., Michalska, K., *Fitoterapia*, (2006), 77, 354.

[34] Takasaki, M., Konoshima, T., Tokuda, H., Masuda, K., Arai, Y., Shiojima, K., Ageta, Hiroyuki., *Biol. Pharm. Bull.*(1999), 22, 606.

[35] Seaman, F. C., *Bot. Rev.,* (1982),48, 123.

[36] Kupchan, S.M., Giacobbe, T.J., Krull, I.S., Thomas, A.M., Eakin, M.A., Fessier, D.C., *J. Org. Chem.* (1970), 35, 3539.

[37] Picman, A.K., *Biochem. Syst. Ecol.*, (1986), 14, 255.

[38] Zdero, C., Bohlmann, F., *Plant Syst. Evol.*, (1990), 171, 1.

[39] Kisiel, W., Barszcz, B., *Fitoterapia*, (2000),71, 269.

[40] EL-Masry, S., Ghazy, N. M., Zdero, K., Bohlmann, F., *Phytochemistry* (1984), 23, 183.

[41] Michalska, K., Kisiel, W., *Biochem. System. Ecol.,* (2007),35, 714.

[42] Takeda, Y., Musuda, T., Morikawa, H., Ayabe, H., Hirata, E., Shinzato, T., Aramato, M., Otsuka, H., *Phytochemistry* (2005),66, 727.

[43] Khalil, A.T., El-Fattah, H.A., Mansour, E.S., *Planta Med.,* (1991), 57, 190.

[44] Nishimura, K., Miyase, T., Ueno, A., Noro, T., Kuroyanagi, M., Fukushima, S., *Phytochemistry* (1986), 25, 2375.

[45] Wang, X.-X., Lin, C.-J., Jia, Z.-J., *Planta Med.*, (2006),72, 764.

[46] Prabhu, K.R., Venkateswarlu, V., *J. Indian Chem. Soc.*, (1969), 46, 176.

[47] Gupta, M. M., Verma, R.K., Singh, S.C., *Fitoterapia*(1989), 60, 476.

[48] Takasaki, M., Konoshima, T., Tokuda, H., Masuda, K., Arai, Y, Shiolima, K., Ageta, H., *BioI. Pharm. Bull.*, (1999), 22, 606.

[49] Giner, R.M., Diaz, J., Manez, S., Recio, M.C., Soriano, C., Rios, J.L., *Biochem. System. Ecol.*, (1992),20, 187.

[50] Belboukhari, N., Cheriti, A., *Pak. J. Biol. Sci.*, (2006),9, 2930.

[51] Bitam, F., Ciavatta, M.L., Manzo, E., Dibi, A., Gavagnin, M., Phytochemistry (2008), 69, 2984.

[52] a) Cheriti A., Saad A., Belboukhari N., Ghezali S.,*Chem. Nat. Prod.*,(2006), 42(3), 360. b) Belboukhari N., Merzoug Z., Cheriti A., Sekkoum K., Yakoubi M., *PhytoChem & BioSub Journal*, (2013), 7(1), 14.

[53] Behari, M., Gupta, R., *Indian. J. Chem. Section B*, (1980), 19B (10), 926.

[54] a) Majumder, P.L., Laha, S., *J. Indian. Chem. Soc.*, (1982), 59, 881. b) Hook, F., Behari, M., Gupta, R., Mutsumoto, T., *Indian Drugs* (1984),21, 366.

[55] Zaheer, A., Dildar, A., Abdul, M., *Magn. Reson. Chem.*, (2006), 44, 717.

[56] Saleem M., Parveen S., Riaz N., Nawaz Tahir M., Ashraf M., Afzal I., Shaiq Ali M., Malik A., Jabbar A., *Phytochemistry Let.,* (2012),5, 793.

[57] a) Abdel-Fattah, H., Zaghloul, A.M., Halim, A.F., Waight, E.S., *Egypt. J. Pharm. Sci.*, (1990), 31, 81. b) Gherraf, N., El-Bassuony, A.A.,

Zellagui, A., Rhouati, S., Ahmed, A.A., Ouahrani, M.R., *Asian. J. Chem*., (2006), 18, 2348.

[58] Zellagui A., Gherraf N., Ladjel S.,Hameurlaine S., *Org. Med. Chem. Let*., (2012), 2, 2.

In: Terpenoids and Squalene
Editor: Alanna R. Bates

ISBN: 978-1-63463-656-8

Chapter 4

SQUALENE: MOLECULAR PROPERTIES, BIOLOGICAL IMPORTANCE AND APPLICATIONS

***Ghayth Rigane*[1,2,*] *and Ridha Ben Salem*[1]**

[1]Laboratoire de Chimie Organique-Physique UR11ES74, Faculté des Sciences de Sfax, Département de Chimie, Université de Sfax, Tunisie
[2]Département de Physique-Chimie, Faculté Des Sciences et Techniques de Sidi Bouzid, Université de Kairouan, Tunisie

ABSTRACT

Squalene is a natural lipid belonging to the terpenoid family and a precursor of cholesterol biosynthesis. Squalene has attracted increasing attention during the past few years due to their biological activities and natural abundance and is potential targets for the food and pharmaceutical industries. Here we summarize the recent knowledge of squalene biochemistry, its molecular properties, and its physiological effects. The analytical techniques used to identify and quantify squalene are also reviewed.

Keywords: Squalene, antioxidant, terpenoid, isoprenoid, antimicrobial

---

* Corresponding author: Dr. Ghayth Rigane. Laboratoire de Chimie Organique-Physique UR11ES74, Faculté des Sciences de Sfax, Département de Chimie, B.P « 1171 » 3038, Sfax, Université de Sfax, Tunisie. E-mail: gaith.rigane@yahoo.fr.

1. INTRODUCTION

Terpenes (or terpenoids) are a group of molecules with extraordinarily diverse chemistry, structure, and function. Given the many ways in which the basic isoprenoid pattern can be assembled, and the different selection pressures under which organisms have evolved, it is not surprising that so many terpenes exist in nature (Gershenzon et al., 2007; Reddy et al., 2009).

In fact, more than 30,000 terpenoid compounds have been characterized to date, constituting the largest group of natural products (Penuelas et al., 2005). A terpene is classified according to the number of its isoprene units (a) and carbon atoms (b) and is identified by the notation a:b, i.e., monoterpenes (2:10), sesquiterpenes (3:15), diterpenes (4:20), sesterpenes (5:25), triterpenes (6:30), carotenoids (8:40), and rubber (>100: >500).

Squalene (2,6,10,15,19,23-hexamethyl-6,6,10,14,18,20-tetracosahexane), a polyunsaturated triterpene containing six isoprene units (Figure 1) with a formula $C_{30}H_{50}$ (Aguilera et al., 2005). It has been identified in the shark liver extract by Dr. Tsujimoto in 1903, hence it received its name (*Squaluss* spp.).

It is widely present in nature: in olive oil, palm oil, wheat-germ oil, amaranth oil, and rice bran oil. Squalene is an oily liquid with low viscosity. When purified, it becomes very pale yellow to colorless, and with almost no odor.

In addition to cosmeceutical studies, squalene is regarded as an important compound for chemoprotective activities as well as nutraceutical for maintaining health under toxic exposure (Das et al., 2003).

Fully natural occurrence and studied beneficial properties of squalene promote this aliphatic hydrocarbon for leading significant improvements on the way to develop nutraceuticals which can help people to maintain a quality and healthy life in today's nonnatural trended life.

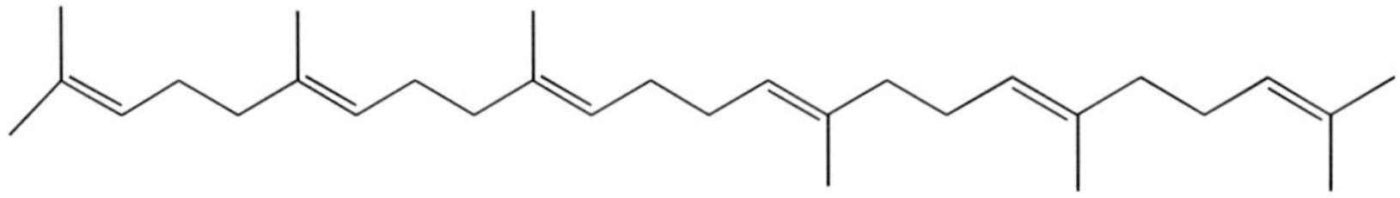

Figure 1. Molecule structure of Squalene.

In this review, we will describe biochemical and biophysical properties of squalene. In addition, we will discuss the role of squalene in olive oil stability.

Finally, we will briefly describe the analytical techniques used to identify and quantify squalene and discuss new developments to use squalene in nutrition, pharmacy, medicine, and cosmetics.

2. Molecular Properties

Squalene, a well known polyunsaturated triterpene, contains six isoprene units. Structurally similar to beta-carotene, it has considered as an intermediate metabolite in the synthesis of cholesterol. The basic skeletal structure of all steroids is formed by the condensation of six prenyl groups.

As it was mentioned in several studies, squalene was known as a liquid with pleasant, bland taste. These data underline the strong hydrophobic nature of this molecule. As it can be seen from Figure 1, double bonds allow squalene to occur in several conformations, e.g., in a symmetric, stretched, or coiled form (Milla et al., 2002; Spanova and Daum, 2011). Due to its chemical structure, especially the high degree of unsaturation, squalene is not very stable and gets easily oxidized. In a previous study, Bolland and Hughes (1949) studied the squalene autoxidation at 55 °C. They deduced that for each oxidized molecule two oxygen molecules were consumed. They suggested the formation of a diperoxide in four steps (Figure 2).

From the sequence of reactions it is evident that the peroxy radical cyclizes more efficiently than it abstracts a hydrogen atom from another squalene molecule. The determining factor appears to be a steric one, i.e., the particular spacing of the double bonds must be sufficiently favorable for cyclization of the peroxy radical.

In addition, it was mentioned that the products of squalene oxidation remain unchanged over a substantial range of oxygen uptake, so that they may not strongly participate in propagation reactions.

Squalene is reported to be well-known as the major olive oil hydrocarbon and makes up more than 90% of the hydrocarbon fraction ranging from 200 to 7500 mg/kg oil or even higher (800-12000 mg/kg oil). Squalene content depends on many factors such as: olive cultivar (Baccouri et al., 2008; Rigane et al., 2013a), oil extraction technology, and it is dramatically reduced during the process of refining (Psomiadou and Tsimidou, 1999).

Very little is known for the contribution of squalene to the oxidative stability of olive oil or other edible oils and fats (Manzi et al., 1998).

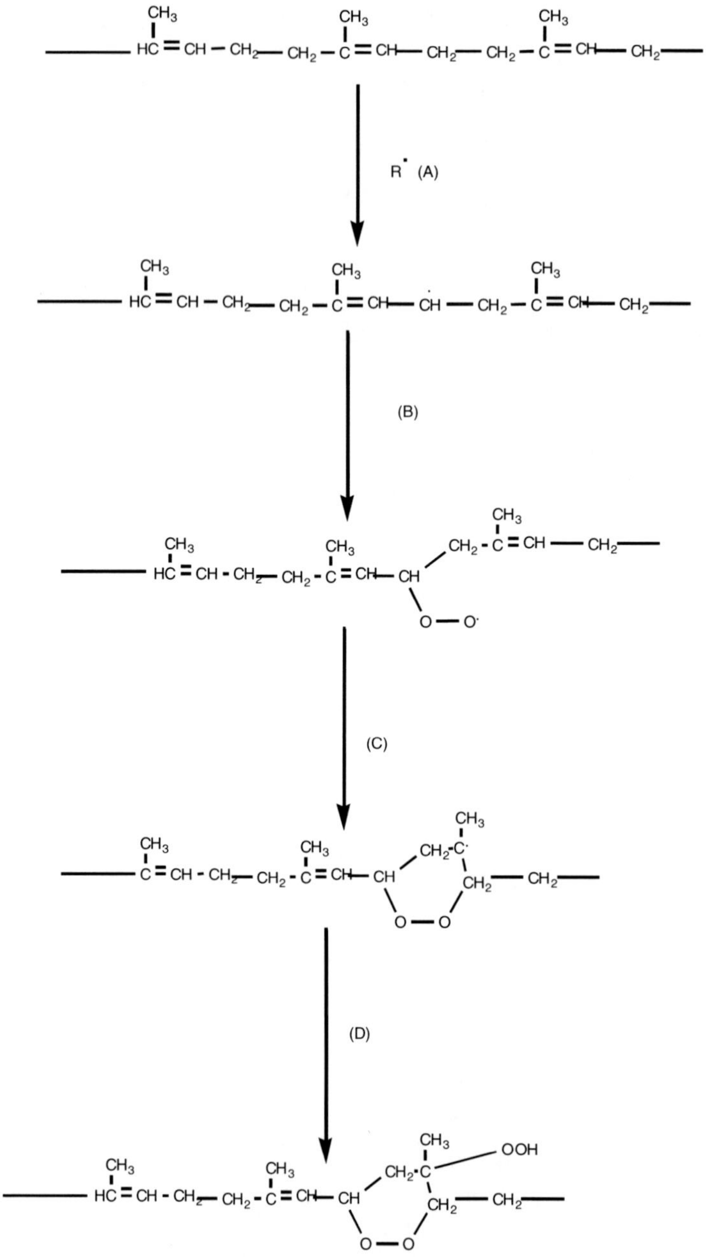

CH3 CH3 CH3
HC = CH – CH2 – CH2 – C = CH – CH2 – CH2 – C = CH – CH2

R˙ (A)

CH3 CH3 CH3
HC = CH – CH2 – CH2 – C = CH – ĊH – CH2 – C = CH – CH2

(B)

CH3 CH3 CH3
HC = CH - CH2 – CH2 - C = CH – CH – CH2 - C = CH – CH2
O – O˙

(C)

CH3 CH3 CH3
C = CH - CH2 – CH2 - C = CH – CH CH2-C˙ CH2 – CH2
O – O

(D)

CH3 CH3 CH3
HC = CH - CH2 – CH2 - C = CH – CH CH2-C OOH CH2 – CH2
O – O

Figure 2. Scheme of squalene oxidation proposed by Bolland and Hughes (1949).

Many research indicated a squalene loss ranged between 26-47 % for virgin olive oils stored at room temperature (~25°C for 6 months of storage) in the dark (Manzi et al., 1998) but not with the low reduction (-10 %) referred to Psomiadou, and Tsimidou (2002). On the other hand, Rastrelli, et al (2002) showed that squalene content in Scalicelle (-45%) and Terzer (-51.5%) olive oil deceased significantly only after 6 months in half-empty bottles, while, after a one year of storage in the filled colorless bottles at diffused lighting, they claimed that diffused lighting does not appear to play a significant role in the squalene degradation (Scalicelle, -19.3%, and Terzera, -23.8%). Nevertheless, Rigane et al. (2013b) observed that squalene could stabilize refined olive oil after enrichment with 800 mg.kg^{-1} of pure squalene. It inhibits thermal deterioration of oil by improving its hydrolytic stability, inhibiting double bond conjugation and reducing the formation of second products of oxidation.

3. Solid Phase Extraction in the Analysis of Squalene

Solid phase extraction (SPE) is only marginally employed in the official EU methods on the characteristics of olive oil. Grigoriadou et al. (2007) had studied the potential of SPE in the official control of virgin olive oil. Thus, silica cartridges were employed for the isolation of constituents that could be eluted earlier than triacylglycerols. The results of the Grigoriadou's study together with those reported by Perez-Camino et al., (2002) can be proved of particular interest in the official control of virgin olive oil and it is expected to attract the attention of officials responsible for that. Using only one silica cartridge, it is possible to prepare fractions quantitatively rich in minor compounds with specific interest either as quality markers or as indices of nutritional value. The procedure was found to be reliable and fast, and allowed recovery of the compounds prior to the analysis of triacylglycerols. The advantages of the above solid phase extraction procedure is in line with current trend in food analysis. Furthermore, the presence of squalene has been allegedly related to various health benefits of virgin olive oil, still concrete proof is limited. Squalene is determined by titrimetric or chromatographic procedures (HPLC, GC, hyphenated chromatographic techniques). The only official method for the determination of squalene is the one proposed by AOAC that involves sample saponification, extraction of the non-saponifiable matter with large quantities of solvents, fractionation through column chromatography, and other treatments just before titration (AOAC, 1999).

4. Squalene: Nutrition, Pharmacy, Medicine, and Cosmetics

Squalene is a biochemical precursor of cholesterol and other steroids. The synthesis of squalene begins with the conversion of acetyl coenzyme A to 3-hydroxy-3-methylglutaryl coenzyme A (HMG CoA), followed by reduction of HMG CoA to mevalonate, mediated by 3- hydroxy-3-methylglutaryl coenzyme A reductase (HMG CoA reductase). Mevalonate is then phosphorylated and eventually decarboxylated to form Δ-3-isopentenyl diphosphate, which is the donor molecule for polyprenyl compounds. The successive addition of prenyl functional groups results in the formation of farnesyl diphosphate. Two farnesyl diphosphate molecules undergo a reductive coupling to form squalene (Kelly, 1999).

In addition, this molecule, i.e: squalene, being unusually high in olive oil, has been the focus of a number of studies, both in humans and laboratory animals. Dietary squalene was shown to inhibit the activity of HMG CoA reductase, the key regulatory enzyme in the cholesterol synthesis pathway in the rat (Strandberg et al., 1989). Studies on dietary squalene supplementation in humans, on the other hand, gave inconsistent results as to its effect on serum cholesterol levels. Preliminary short-term studies in elderly subjects showed inconsistent changes in serum cholesterol levels (Miettinen et al., 1986). Strandberg et al, (1990) observed no change in serum cholesterol levels after 7-30 days of squalene feeding at 900 mg/day. In the some way, Miettinen and Vanhanen (1994) reported that adding 1 g of squalene in rapeseed oil for 9 weeks caused a net increase in serum cholesterol levels.

Therefore, further studies are required to clarity the potential hypocholesterolemic or hypercholesterolemic effect of dietary squalene. Thus, in addition to being synthesized within cells, it is consumed as an integral part of the human diet. In humans, squalene is synthesized in the liver and the skin, transported in the blood by very low density lipoproteins (VLDL) and low density lipoproteins (LDL), and secreted in large quantities by the sebaceous glands (Stewart, 1982; Koivisto and Miettinen, 1988).

Squalene is suggested to show not only antitumor but also chemoprotective effects.

Storm et al. (1993) have reported that mice were administrated by 2% squalene for 14 days before and 30 days after a lethal whole-body g-irradiation. Results showed that white cell counts in squalene-treated groups were considerably high when compared to nontreated control groups. Also, white cell protection was observed along a prolonged survival. Squalene treatment was suggested to exhibit cellular and systemic radioprotection as a result of this research. In a related research, Nakagawa et al. (1985) have reported the interaction between squalene and anticancer agents. They mentioned that a potentiating effect of squalene was observed for Adriamycin, 5-fluorouracil, bleomycin, and cis-diamminedichloroplatinum. Squalene addition enhanced the cytotoxicity and antitumor activity. In similar studies, it has been shown that squalene treatment before and/or during anticancer treatment effectively enhanced the inhibition of chemically induced skin tumorigenesis and regression of some already existing tumors in animal models. All results indicate that squalene makes a big contribution to enhance tumor growth inhibition and tumor suppression by anticancer treatments as well as by acting as a successful chemoprotective agent.

During recent years, squalene was found to be active for protection against various carcinogens. Accordingly, the correlation between high amounts of squalene and absence of cancer in shark species ought to be suggested. There are reports which suggest partial and possible relationship between low incidence of cancer and consuming high olive oil product which also includes high amounts of squalene (Owen et al., 2000, 2004).

On the other hand, many of researchers and clinicians are worried about women who consume alcohol in excess during pregnancy which could be diagnosed with fetal alcohol syndrome (FAS), to be associated with growth retardation, disorders of the central nervous system (CNS), a characteristic pattern of facial anomalies and more or less severe deficits involving cognitive and behavioural disorders. Fortunately, most women reduce or stop their alcohol consumption as soon as they realize they are pregnant; in some cases, however, some injury to the fetal brain may have already occurred (Jones and Smith, 1973; Maier and West, 2001). For this purpose, and taking into account the harmful effects of alcohol on health wich could be mostly by reactive oxygen species, especially by some oxygen and other free radicals (Mantle and Preedy, 1999; Zloch, 1994) and that the retina and optic nerve, Aguilera et al. (2005) have studied the protective role of squalene in alcohol-induced damage during the chick retinal development. They demonstrated that squalene reduces the deleterious effects caused by alcohol on the lipid composition and the structure of the retina, squalene could act as a naturally occurring agent for the prevention of damage caused by abusive alcohol ingestion during pregnancy.

Recently, Nowicki and Bara ska-Rybak (2007) observed a significant protection against bacterial and fungal infections by shark liver oil treatment which contains mostly squalene and alkylglycerol. Further, this treatment showed improved effects on xerosis and skin lesion-induced atopic dermatitis. This antibacterial and antifungal effect could be accounted for the high squalene- including composition of the shark liver oil.

Applications of squalene have been recently reviewed in some detail (Kohno et al., 1995; Newmark, 1999). Warleta et al. (2010) reported that antiradical activities of squalene have been measured by 2,2-diphenyl-1-picrylhydrazyl (DPPH), 2,2-azino-bis(3-ethylbenzothiazoline- 6-sulphonic acid) (ABTS), and oxygen radical absorbance capacity assays. According to assay results, radical scavenging capacity of squalene has been reported to be significantly important to be an efficient free radical scavenger. Interestingly, they concluded that the squalene antioxidant selectivity depends either on the ‘‘glutathione paradox’’, where squalene increases the amount of glutathione in normal cells (Das et al., 2003); on differences in squalene uptake, utilization, and accumulation (Das et al., 2008) ; or on deregulation of antioxidant systems in tumor cells (Klaunig et al., 2004).

Squalene is frequently used in the preparation of stable emulsions as either the main ingredient or secondary oil (Tadrosa et al., 2004; Fox et al., 2004; Whittenton et al., 2008). Squalene emulsions have been used for various applications, especially the delivery of vaccines, drugs, and other medicinal substances. As an example, squalene together with the detergents Tween 80 and Span 85 forms the adjuvant MF59 (Novartis), an oil-in-water microemulsion approved for human use (Tagliabue et al., 2008; Schultze et al., 2008). MF59 has been shown to be a potent and safe adjuvant with several vaccines, e.g., against hepatitis B and C, herpes simplex virus, HIV-1, and influenza (vaccine Fluad).

Conclusion

In this review article, we summarized our recent knowledge about squalene, an isoprenoid lipid and intermediate of sterol synthesis. Squalene, which can be obtained by natural sources, exists readily in human body and yet to be examined in detail for its biological role. In addition, squalene offers unique characteristics in relation to other dietary lipid constituents and several of its pharmacological functions are discussed here. Although to date squalene is indicated only for vaccine delivery and used for cosmetic purposes, the diverse applications demonstrated at the preclinical stage in drug and gene delivery clearly suggest that this natural triterpene has a very bright future in disease management and therapy. Squalene appears to influence several biochemical and physiological activities which are intriguing for the treatment of cancer. In summary, squalene can be regarded as a versatile molecule which may become even more useful for applications in the future.

ACKNOWLEDGMENTS

The authors thank the Tunisian Ministry of Higher Education and Scientific Research for financial support and are grateful to Professor Mohamed Rigane for useful discussions about the English.

REFERENCES

Aguilera, Y., Dorado, M. E., Prada, F. A., Martınez, J. J., Quesad, A., and Ruiz-Gutierrez, V. (2005). The protective role of squalene in alcohol damage in the chick embryo retina. *Experimental Eye Research*, 80, 535-543.

AOAC (1999). Squalene in oils and fats. Titrimetric method. Official method. *AOAC*, p. 943.04.

Baccouri, O., Guerfel, M., Baccouri, B., Cerretani, L., Bendini, A., Lercker, G., Zarrouk, M., and Ben Miled, D. D. (2008). Chemical composition and oxidative stability of Tunisian monovarietal virgin olive oils with regard to fruit ripening. *Food Chemistry*, 109, 743-754.

Bolland, J. L. and Hughes, H. (1949). The primary thermal oxidation product of squalene. *Journal of the Chemical Society*, 492-497.

Das, B., Antoon, R., Tsuchida, R., Lotfi, S., et al. (2008). Squalene selectively protects mouse bone marrow progenitors against cisplatin and carboplatin-induced cytotoxicity in vivo without protecting tumor growth. *Neoplasia*, 10, 1105-1U53.

Das, B., Yeger, H., Baruchel, H., Freedman, M., Koren, G., and Baruchel, S. (2003). In vitro cytoprotective activity of squalene on a bone marrow versus neuroblastoma model of cisplatin-induced toxicity: Implications in cancer chemotherapy. *European Journal of Cancer*, 39, 2556-2565.

Fox, C. B., Anderson, R. C., Dutill, T. S., Goto, Y., Reed, S. G., and Vedvick, T. S. (2008). Monitoring the effects of component structure and source on formulation stability and adjuvant activity of oil-in-water emulsions, Colloids Surf. *B Biointerfaces*, 65, 98-105.

Gershenzon, J. and Dudareva, N. (2007). The function of terpene natural products in the natural world. *Nature Chemical Biology*, 3, 408-414.

Grigoriadou, D., Androulaki, A., Psomiadou, E., and Tsimidou, M. Z. (2007). Solid phase extraction in the analysis of squalene and tocopherols in olive oil. *Food Chemistry,* 105,675-680.

Jones, K. L. and Smith, D. W. (1973). Recognition of the fetal alcohol syndrome in early infancy. *Lancet*, 2, 999-1001.

Kelly, G. S. (1999). Squalene and its potential clinical uses. *Alternative medicine review*, 4 29-36.

Klaunig, J. E. and Kamendulis, L. M. (2004). The role of oxidative stress in carcinogenesis. *Annual Review of Pharmacology and Toxicology*, 44, 239-267.

Kohno, Y., Egawa, Y., Itoh, S., Nagaoka, S., Takahashi, M., and Mukai, K. (1995). Kinetic study of quenching reaction of singlet oxygen and scavenging reaction of free radical by squalene in n-butanol. *Biochimica et Biophysica Acta*, 1256, 52-56.

Koivisto, P. V. I. and Miettinen, T. A. (1988). Increased amount of cholesterol precursors in lipoproteins after ileal exclusion, *Lipids*, 23, 993-996.

Maier, S. E. and West, J. R. (2001). Drinking patterns and alcohol-related birth defects. *Alcohol Research Health*, 25, 168-174.

Mantle, D. and Preedy, V. R. (1999). Free radicals as mediators of alcohol toxicity. *Adverse drug reactions and toxicological reviews*, 18, 235-252.

Manzi, P., Panfili, G., Esti, M., and Pizzoferrato, L. (1998). Natural antioxidants in the unsaponifiable fraction of virgin olive oils from different cultivars. *Journal of the Science and Food Agriculture.* 77, 115-120, (1998).

Miettinen, T. A. and Kesaniemi, Y. A. (1986). Cholesterol balance and lipoprotein metabolism in man. In: Grundy, S. M., ed. Bile acids and atherosclerosis. New York: Raven Press. 113-115.

Miettinen, T. A. and Vanhanen, H. (1994). Serum concentration and metabolism of cholesterol during rapeseed oil and squalene feeding. *The American Journal of Clinical Nutrition*, 59, 356-363.

Milla, P., Athenstaedt, K., Viola, F., Oliaro-Bosso, S., et al., (2002). Yeast oxidosqualene cyclase (Erg7p) is a major component of lipid particles. *Journal of Biological Chemistry*, 277, 2406-2412.

Newmark, H. L. (1999). Squalene, olive oil, and cancer risk. Review and hypothesis. *Annals of the New York Academy of Sciences*, 889, 193-203.

Nowicki, R. and Bara ska-Rybak, W. (2007). Shark liver oil as a supporting therapy in atopic dermatitis. *Polski Merkuriusz Lekarski*, 22, 312-313.

Owen, R., Haubner, R., Wurtele, G., Hull, W., Spiegelhalder, B., and Bartsch, H. (2004). Olives and olive oil in cancer prevention. *European Journal of Cancer Prevention*, 13, 319-326.

Owen, R. W., Giacosa, A., Hull, W. E., Haubner, R., Wurtele, G., Spiegelhalder, B., and Bartsch, H. (2000). Olive-oil consumption and health: The possible role of antioxidants. *The Lancet Oncology*, 1, 107-112.

Penuelas, J. and Munné-Bosch, S. (2005). Isoprenoids: an evolutionary pool for photoprotection. *Trends in Plant Science*, 10, 166-169.

Perez-Camino, M. C., Moreda, W., Mateos, R., and Cert, A. (2002). Determination of esters of fatty acids with low molecular weight alcohols in olive oils. *Journal of Agricultural and Food Chemistry*, 50, 4721-4725.

Psomiadou, E. and Tsimidou, M. (1999). On the Role of Squalene in Olive Oil Stability. *Journal of Agriculture and Food Chemistry*. 47, 4025-4032.

Psomiadou, E. and Tsimidou, M. (2002). Stability of virgin olive oil. 1. Autoxidation studies. *Journal of Agriculture and Food Chemistry,* 50, 716-721.

Rastrelli, L., Passi, S., Ippolito, F., Vacca, G., and De Simone, F. (2002). Rate of Degradation of α-Tocopherol, Squalene, Phenolics, and Polyunsaturated Fatty Acids in Olive Oil during Different Storage Conditions. *Journal of Agriculture and Food Chemistry*, 50, 5566-5570.

Reddy, L. H. and Couvreur, P. (2009). Squalene: A natural triterpene for use in disease management and therapy. *Advanced Drug Delivery Reviews*, 61, 1412-1426.

Rigane, G., Bouaziz, M., Sayadi, S., and Ben Salem, R. (2013b). Effect of storage on refined olive oil composition: Stabilization by addition of chlorophyll pigments and squalene. *Journal of Oleo Science*, 981-987.

Rigane, G., Boukhris, M., Bouaziz, M., Sayadi, S., and Ben Salem, R. (2013a). Analytical evaluation of two monovarietal virgin olive oils cultivated in South of Tunisia: Jemri-Bouchouka and Chemlali-Tataouin cultivars. *Journal of the Science of Food and Agriculture,* 93, 1242-1248.

Schultze, V., D'Agosto, V., Wack, A., Novicki, D., et al., (2008). Safety of MF59 adjuvant. *Vaccine*, 26, 3209-3222.

Spanova, M. and Gunther, D. (2011). Squalene – biochemistry, molecular biology, process biotechnology, and applications. *European Journal of Lipid Science and Technology*, 113, 1299-1320.

Stewart, M. E. (1992). Sebaceous gland lipids, *Seminars in dermatology*, 11, 100-105.

Storm, H. M., Oh, S. Y., Kimler, B. F., and Norton, S. (1993). Radioprotection of mice by dietary squalene. *Lipids*, 28, 555-559.

Strandberg, T. E., Silvis, R. S. and Miettinen, T. A. (1989). Effect of cholestyramine and squalene feeding on hepatic and serum plant sterols in the rat. *Lipids*, 24, 705-708.

Strandberg, T. E., Silvis, R. S. and Miettinen, T. A. (1990). Metabolic variables of cholesterol during squalene feeding in humans: comparison with cholestyramine treatment. *The Journal of Lipid Research*, 31, 1637-1643.

Tadros, T., Izquierdo P., Esquen, J., and Solans, C. (2004). Formation and stability of nanoemulsions, *Advances in Colloid and Interface Science*, 108-109, 303-318.

Tagliabue, A. and Rappuoli, R. (2008). Vaccine adjuvants - The dream becomes real. *Human Vaccines and Immunotherapeutics*, 4, 347-349.

Warleta, F., Campos, M., Allouche, Y., Sanchez-Quesada, C., Ruiz-Mora, J., Beltran, G., and Gaforio, J. J. (2010). Squalene protects against oxidative DNA damage in mcf10a human mammary epithelial cells but not in mcf7 and mda-mb-231 human breast cancer cells. *Food and Chemical Toxicology*, 48, 1092-1100.

Whittenton J., S. Harendra, R. Pitchumani, K. Mohanty, C., and Vipulanandan, S. (2008). Thevananther, Evaluation of asymmetric liposomal nanoparticles for encapsulation of polynucleotides, *Langmuir*, 24, 8533-8540.

Zloch, Z. (1994). Temporal changes of the lipid peroxidation in rats after acute intoxication by ethanol. *Z. Naturforsch*, 49, 359-362.

INDEX

A

B

D

E

F

G

H

I

U

V

W

Y